Lecture Notes in Mathematics

Edited by A. Dold and B. Eckmann

700

Module Theory

Papers and Problems from
The Special Session Sponsored by
The American Mathematical Society at
The University of Washington

Proceedings, Seattle, August 15–18, 1977

Edited by Carl Faith and Sylvia Wiegand

Springer-Verlag
Berlin Heidelberg New York 1979

Editors
Carl Faith
Department of Mathematics
Rutgers,
The State University
New Brunswick, New Jersey 08903/USA

Sylvia Wiegand
Department of Mathematics
University of Nebraska
Lincoln, Nebraska 68588/USA

AMS Subject Classifications (1970): 13A15, 13B20, 13B99, 13C10, 13C15, 13D05, 13F05, 13F15, 13F20, 13H10, 16A08, 16A34, 16A52, 16A62, 16A64, 16A66, 18GXX.

ISBN 3-540-09107-6 Springer-Verlag Berlin Heidelberg New York
ISBN 0-387-09107-6 Springer-Verlag New York Heidelberg Berlin

Printed in Germany

Printing and binding: Beltz Offsetdruck, Hemsbach/Bergstr.
2141/3140-543210

"I am going to hang up the gloves next year"
(Nathan Jacobson)

DEDICATION

The contributors and particpants of the Special Session dedicate this volume to Nathan Jacobson in admiration , and gratitude for showing us what to do in the ring (theory).

PREFACE

The editors wish to thank the participants and contributors for their splendid cooperation, and for their joie de vivre which made the Special Session so much fun.

The senior editor has grouped the contributed papers along ideological lines whenever possible, although, like much ideology, these are far-fetched in many cases. It therefore would serve no useful purpose to expose this here, but the reading is certainly better served this way than the old ABC way. One ought to mention that Dr. J. T. Stafford appeared first in the program, as a guest of the Society introduced by the senior editor, and that Dr. Warfield's paper just has to follow Stafford's. Et cetera.

LIST OF PARTICIPANTS AND CONTRIBUTORS

G. Azumaya
Indiana University
Bloomington, IN 47401

J. Beachy
Northern Illinois Univ.
DeKalb, IL 60115

A. K. Boyle
University of Wisconsin
Milwaukee, WI 53706

V. Camillo
University of Iowa
Iowa City, IA 52240

C. Faith
Rutgers University
New Brunswick, NJ 08903

K. R. Fuller
University of Iowa
Iowa City, IA 52240

F. Hansen
Universität Bochum
D - 4630 Bochum

M. Hochster
University of Michigan
Ann Arbor, MI 48109

L. S. Levy
University of Wisconsin
Madison, WI 53706

E. Matlis
Northwestern University
Evanston, IL 60201

S. Mohamed
Kuwait University
Kuwait

B. L. Osofsky
Rutgers University
New Brunswick, NJ 08903

Z. Papp
George Mason University
Fairfax, VA 22030

M. Smith
University of Texas
Austin, TX 78712

J. T. Stafford
Brandeis University
Waltham, MA 02154

M. L. Teply
University of Florida
Gainesville, FL 32611

R. Warfield, Jr.
University of Washington
Seattle, WA 98195

R. Wiegand
University of Nebraska
Lincoln, NB 68588

S. Wiegand
University of Nebraska
Lincoln, NB 68588

TABLE OF CONTENTS

Papers

Problems

PAPERS

CANCELLATION FOR NONPROJECTIVE MODULES

J. T. Stafford
Brandeis University
Waltham, Massachusetts 02154

In [7] Serre showed that, given a commutative Noetherian ring R and a projective, finitely generated R-module M with $f - rk(M) \geq \dim(\max(R)) + 1$, then $M \cong M' \oplus R$. When Bass considered this result in [1], he was able to remove the projectivity condition. He was further able to show that, if $M \oplus R \cong N \oplus R$, then $M \cong N$. However for this cancellation theorem he still had to require that M had a "large" projective direct summand. Thus the obvious question remains as to whether this second result still holds without the projectivity condition. In this paper we answer this question affirmatively. This comes as a corollary of the main results of this paper, where we prove that the above two theorems hold for modules over fully bounded J-Noetherian rings.

Of course, to do this we need a definition of rank that requires no localization. For this purpose we use the $r - rk$ of [8] (which for a finitely generated module over a commutative Noetherian ring is equivalent to $f - rk$). In fact the proofs given here closely follow those given in [8], where versions of the above two theorems were proved for noncommutative Noetherian rings. However the results given there used, for the dimension on the ring, the Krull dimension of Rentschler and Gabriel [6], which, of course, is in general larger than $\dim(\max(R))$. However it was claimed that the methods used in [8] would answer the question posed above, and this paper can be considered as a substantiation of that assertion.

Throughout this paper all rings will contain an identity and all modules will be unitary.

§1. Notation and Preliminary Results.

The results of this paper hold for a more general class of rings than Noetherian rings and we start by defining that class. The notation comes from [3]. Let R be a ring and I an ideal of R. Then I is called a J-ideal (respectively a J-prime) if I is an ideal (respectively a prime ideal) that is the intersection of the maximal ideals containing it. The ring is called J-Noetherian if it has the ascending chain condition on J-ideals. Define J-dim R to be the maximal length n of chains of J-primes,

$$J_0 \subsetneqq J_1 \subsetneqq \cdots \subsetneqq J_n \subsetneqq R.$$

If R is commutative then these concepts coincide with the ones used by Bass. In particular, R is J-Noetherian if and only if the max spectrum of R is Noetherian, in which case J-dim R = dim(max(R)).

As mentioned in the introduction, we will prove the results of this paper for more than just commutative rings. Define a prime ring R to be left bounded if any essential left ideal contains a non-zero ideal. Define a ring R to be fully left bounded Goldie if every prime factor ring is left bounded and left Goldie. For example, commutative rings, PI rings, and FBN rings are fully left bounded Goldie. One of the crutial facts that we need about these rings is the following.

Lemma 1.1: Let R be a fully left bounded Goldie ring. Then any left primitive factor ring of R is a simple Artinian ring.

Proof: We may suppose that R is left primitive. So there exists a maximal left ideal M that does not contain an ideal. Thus M is not essential, and there exists a left ideal K such that $M \cap K = 0$. Clearly K is simple. So R is a simple Artinian ring by [4, 3.50].

Finally we need to define two concepts of rank. If R is a ring, finitely generated over its center C and M is an R-module, the usual definition of rank is given by

$f - rk(M) \geq s$ if, given any maximal ideal P of C then
$$M_P \cong R_P^{(s)} \oplus M' \text{ for some module } M' = M'(P).$$

However in this paper we will use a different definition of rank. Given a ring R and a module M define;

$r - rk(M) \geq s$ if, given any prime ideal P of R and elements $\alpha_1, \ldots, \alpha_{s-1} \in M$, then there exists $\alpha_s \in M$ and $\theta \in \mathrm{Hom}(M, R)$ such that $\theta(\alpha_i) \in P$ for $1 \leq i \leq s - 1$ yet $\theta(\alpha_s) \in \mathcal{C}(P)$. Here $\mathcal{C}(P)$ is the set of elements of R that become regular in R/P.

The reason for this definition is two-fold. First, for a commutative ring it enables us to prove the cancellation theorem without projectivity conditions on the module. Second, it enables us to prove both Serre's theorem and the Cancellation Theorem for modules over fully left bounded Goldie rings, where $f - rk$ will not usually be defined. However when the ring is a finite module over its center, the two concepts coincide.

Proposition 1.2: Let C be a commutative Noetherian ring and R a finite C-algebra. If M is a finitely generated R-module then

$$f - rk(M) = r - rk(M).$$

Proof: The case $R = C$ is given in [8, Proposition 2.6]. The general case is a fairly easy adaption of that case. See, for example, the proof of [2, Theorem 7.9].

§2. Serre's Theorem.

We start by proving Serre's Theorem; i.e. that "big" modules have free direct summands. Unfortunately, even for commutative rings, we need to prove this in order to be able to prove the Cancellation Theorem, as to prove the latter result we need the proof rather than the result of Serre's Theorem.

Let R be a ring and M a right R-module. Given $\alpha_1, \ldots, \alpha_r \in M$ define

$$U_r = U(\alpha_1, \ldots, \alpha_r) = \{(f(\alpha_1), \ldots, f(\alpha_r)) : f \in \text{Hom}\ (M, R)\} \subseteq {}_R R^{(r)}.$$

Given a prime P, say that U_r is good at P if $R^{(r)}/U_r + P^{(r)}$ is a torsion left R/P-module. U_r is bad at P if it is not good. Define e_i to be the element of $R^{(r)}$ with a 1 in the $i^{\underline{th}}$ copy of R and zeroes elsewhere. For the next two lemmas we fix an R-module M and elements $\alpha_1, \ldots, \alpha_r \in M$.

Lemma 2.1: Let R be a ring with a prime ideal P such that R/P is left Goldie. Then the following three conditions are equivalent:

i) U_r is good at P,

ii) for each $i \leq r$ there exists $c \in \mathcal{C}(P)$ such that $e_i c \in U_r + P^{(r)}$,

iii) U_{r-1} is good at P and there exists $\theta \in \mathrm{Hom}\,(M, R)$ such that $\theta(\alpha_i) \in P$ for $i \leq i \leq r - 1$ yet $\theta(\alpha_r) \in \mathcal{C}(P)$.

Proof: This is clear.

Lemma 2.2: Let R be a fully left bounded Goldie, J-Noetherian ring. Let $\{P_i : i \in I\}$ be the J-primes minimal such that U_r is bad at each P_i. Then I is a finite set.

Proof: Suppose I is infinite and consider an ideal J that is equal to the intersection of infinitely many of the P_i's. Since each P_i is a J-ideal, so is J. Thus we may choose J to be maximal among those ideals that are the intersections of infinitely many P_i's. Standard arguments show that J is a J-prime and hence U_r is good at J. So by Lemma 2.1(ii) and the definition of fully bounded, $U_r + J^{(r)}$ contains, for each $i \leq r$, a submodule of the form $e_i T_i$ where T_i is an ideal strictly containing J. Now suppose that $\{P_i : i \in I'\}$ is the subset of $\{P_i : i \in I\}$ consisting of those primes containing J. Then J is equal to the intersection of any infinite subset of $\{P_i : i \in I'\}$. So if some T_j is contained in infinitely many of the P_i's for $i \in I'$, then

$T_j \subset J$, a contradiction. Thus there exists $i \in I'$ such that, for $1 \leq j \leq r$, $T_j \not\subseteq P_i$. But this implies that U_r is good at P_i, a contradiction.

We will prove Serre's Theorem by considering an inductive statement, which will also be used in the proof of the Cancellation Theorem. Let R be a ring and M a right R-module. Then, for $0 \leq r \leq \infty$ let $K(r)$ be the following statement:

$K(r)$ Let $\alpha_1, \ldots, \alpha_s \in M$ with $0 \leq s \leq r$ be such that, given any J-prime P with J-dim $R/P \geq r$, then U_s is good at P. Let $\{P_j : j \in \Omega\}$ be a finite set of J-primes of R with J-dim $(R/P_j) < r$ for all $j \in \Omega$. Then there exists $\alpha_{s+1} \in M$ such that:

a) Given any J-prime P such that J-dim $(R/P) \geq r$, then U_{s+1} is good at P.

b) For all $j \in \Omega$, there exists $\theta_j \in \text{Hom}(M, R)$ such that $\theta_j(\alpha_i) \in P_j$ for $1 \leq i \leq s$ yet $\theta_j(\alpha_{s+1}) \in \mathcal{C}(P_j)$.

The next theorem shows that these statements do hold for large modules over a fully bounded J-Noetherian ring and, as a corollary, we obtain Serre's Theorem.

<u>Theorem 2.3</u>: Let R be a fully left bounded Goldie, J-Noetherian ring with J-dim $R = n$. Let M be a right R-module with r - rk $M \geq n + 1$. Then $K(r)$ holds for M for $0 \leq r \leq n$.

<u>Remark 2.4</u>: It should be noted that the left bounded condition is only used to prove Lemmas 1.1 and 2.2. Thus this theorem and the remaining

results of this paper will still hold for any J-Noetherian ring R, all of whose prime factor rings are left Goldie, such that the results of Lemmas 1.1 and 2.2 hold for R.

Proof of Theorem 2.3: We start by showing that $K(n)$ holds. So suppose that the initial hypotheses are as in $K(n)$. Write $\{P_j : j \in \Omega\}$ as $\{P_1, \ldots, P_u\}$ and let $P_{u+1}, \ldots, P_v$ be the J-primes P of R such that J-dim $R/P = n$ (there are only finitely many by Lemma 2.2). Now by Lemma 2.2 iii) it suffices to find $\alpha_{s+1} \in M$ and, for $1 \leq i \leq v$, homomorphisms $\theta_i \in \text{Hom}(M, R)$ such that $\theta_i(\alpha_j) \in P_i$ for $1 \leq j \leq s$ yet $\theta_i(\alpha_{s+1}) \in \mathcal{L}(P_i)$. But this is precisely what is proved by [8, Proposition 3.4 i)]. So $K(n)$ holds.

So suppose that $K(r + 1)$ is true, and that the hypotheses are as in $K(r)$. By $K(r + 1)$ choose α_{s+1} such that:

a) For all J-primes P with J-dim $R/P \geq r + 1$, U_{s+1} is good at P.

b) The result of $K(r)$ b) holds for α_{s+1}.

Now let $Q_1, \ldots, Q_u$ be the J-primes Q such that U_{s+1} is bad at Q but J-dim $R/Q = r$. (By the choice of α_{s+1} and Lemma 2.2 there are only finitely many such primes.) By $K(r + 1)$, again, choose $\alpha_{s+2} \in M$ such that:

c) For all J-primes P with J-dim $R/P \geq r + 1$, U_{s+2} is good at P.

d) For each $i \leq u$, there exists $\phi_i \in \text{Hom}(M, R)$ such that $\phi_i(\alpha_j) \in Q_i$ for $i \leq j \leq s + 1$, yet $\phi_i(\alpha_{s+2}) \in \mathcal{L}(Q_i)$.

Next, given $\lambda \in R$ define $U_\lambda = U(\alpha_1, \ldots, \alpha_s, \alpha_{s+1} + \alpha_{s+2}\lambda)$. Given a prime P of R clearly $R^{(s+1)}/U_\lambda + P^{(s+1)}$ is a homomorphic image of $R^{(s+2)}/U_{s+2} + P^{(s+2)}$. Thus if U_{s+2} is good at P, so is U_λ. Let $Q_1, \ldots, Q_v$ be the J-primes with J-dim $R/Q_i = r$ such that U_{s+2} is bad at each Q_i (our notation is consistent since, by Lemma 2.1, if U_{s+1} is bad at some Q, then so is U_{s+2}). Note that since J-dim $R/P_j < r$ for each $j \in \Omega$, the sets $\{P_j : j \in \Omega\}$ and $\{Q_1, \ldots, Q_v\}$ have no elements in common. So we can choose

$$\lambda \in \bigcap\{P_j : j \in \Omega\} \cap \bigcap\{Q_i : u + 1 \leq i \leq v\} \cap \bigcap\{ \mathcal{I}(Q_i) : 1 \leq i \leq u\}.$$

We claim that $\alpha'_{s+1} = \alpha_{s+1} + \alpha_{s+2}\lambda$ satisfies the conclusions of $K(r)$. We show this by checking several cases. First, since α_{s+1} was chosen to satisfy $K(r)b$ and $\lambda \in \bigcap\{P_j : j \in \Omega\}$, clearly α'_{s+1} satisfies $K(r)b$. If Q is a J-prime such that $Q \neq Q_i$ for $1 \leq i \leq v$, yet J-dim $R/Q \geq r$, then U_λ is good at Q by the comments of the last paragraph. If $Q = Q_i$ for $u + 1 \leq i \leq v$ then, since U_{s+1} is good at Q and $\lambda \in Q$, U_λ has to be good at Q. Finally suppose that $Q = Q_i$ for $1 \leq i \leq u$. Then since $\lambda \in \mathcal{I}(Q)$, by condition (d) above there exists $\phi \in \mathrm{Hom}\,(M, R)$ such that $\phi(\alpha_i) \in Q$ for $1 \leq i \leq s$ but $\phi(\alpha'_{s+1}) \in \mathcal{I}(Q)$. Thus since U_s is good at Q, U_λ has to be good at Q. Thus α'_{s+1} does satisfy the conclusions of $K(r)$. Induction now completes the proof of the theorem.

<u>Corollary 2.5</u>: Let R be a fully left bounded Goldie, J-Noetherian ring with J-dim $R = n$. Let M be a right R-module with r - rk $M \geq n + 1$. Then $M \cong M' \oplus R$ for some module M'.

Proof: By the theorem, $K(0)$ holds. Thus there exists $\alpha \in M$ such that $U(\alpha)$ is a left ideal of R which is good at every J-prime ideal of R. But, by Lemma 1.1, this is only possible if $U(\alpha) = R$, i.e. if α generates a free direct summand of M.

§3. The Cancellation Theorem.

Following Bass, in order to prove the Cancellation Theorem, we first prove the following theorem.

Theorem 3.1: Let R be a fully left bounded Goldie, J-Noetherian ring with $\text{J-dim } R = n$. Let M be a right R-module with $r - rk(M) \geq n + 1$. Suppose $\alpha \oplus t$ is a unimodular element of $M \oplus R$; that is it generates a free direct summand of $M \oplus R$. Then there exists $\beta \in M$ such that $\alpha + \beta t$ is a unimodular element of M.

Proof: Consider the following inductive statements:

$M(r)$ There exists $\beta \in M$ such that $U(\alpha + \beta t)$ is good at any J-prime P with $\text{J-dim } R/P \geq r$.

We first show that $M(n)$ holds. This is trivial if $n = 0$, so suppose $n > 0$. Let $P_1, \ldots, P_m$ be the minimal J-primes of R. By the Chinese Remainder Theorem and the definition of $r - rk$, there exists $\beta \in M$ such that, for $1 \leq i \leq m$, there exists $\theta_i \in \text{Hom}(M, R)$ such that $\theta_i(\alpha) \in P_i$ yet $\theta_i(\beta) \in \mathcal{C}(P_i)$. Now consider $\alpha + \beta t = \alpha'$. Certainly $\alpha' \oplus t$ is still a unimodular element of $M \oplus R$, so there exists $\phi \in \text{Hom}(M, R)$ such that $1 = \phi(\alpha') + rt$ for some $r \in R$. Thus $1 - rt \in U(\alpha')$. But

$$\theta_i(\alpha') \equiv \theta_i(\alpha + \beta t) \equiv \theta_i(\beta)t \equiv s_i t \bmod P_i,$$

for some $s_i \in \mathcal{C}(P_i)$. Since R/P_i is left Goldie, there exists $s_i' \in \mathcal{C}(P_i)$ and $r_i' \in R$ such that $s_i' r \equiv r_i' s_i \bmod P_i$. Hence $s_i' + P_i = s_i'(1 - rt) + r_i' s_i t + P_i \in U(\alpha') + P_i$. Thus $U(\alpha')$ is good at all the minimal J-primes, which is certainly sufficient.

Now suppose that $M(r + 1)$ holds. Replacing α by $\alpha + \beta t$ for some $\beta \in M$, we may suppose that $U(\alpha)$ is good at all J-primes P with J-dim $R/P \geq r + 1$. Let $Q_1,\ldots,Q_u$ be the J-primes with J-dim $R/Q_i = r$ such that $U(\alpha)$ is bad at each Q_i (there are only finitely many by Lemma 2.2). By $K(r + 1)$ and Theorem 2.3 there exists $\beta \in M$ such that:

a) $U(\alpha, \beta)$ is good at any J-prime P with J-dim $R/P \geq r + 1$.

b) For $1 \leq i \leq u$ there exists $\theta_i \in \mathrm{Hom}(M, R)$ such that $\theta_i(\alpha) \in Q_i$ yet $\theta_i(\beta) \in \mathcal{C}(Q_i)$.

Let $Q_1,\ldots, Q_v$ be the J-primes with J-dim $R/Q_i = r$ such that $U(\alpha, \beta)$ is bad at each Q_i (the notation is consistent since, if $U(\alpha)$ is bad at some Q, then so is $U(\alpha, \beta)$). Now choose

$$\lambda \in \bigcap\{Q_i : u + 1 \leq i \leq v\} \cap \bigcap\{\mathcal{C}(Q_i) : 1 \leq i \leq u\}.$$

Then $\alpha + \beta\lambda t$ satisfies the hypotheses of $M(r)$. (For $Q_1,\ldots, Q_u$ the proof of this follows by the argument used for $M(n)$. For all other J-primes use the argument given at the end of the proof of Theorem 2.3.) Thus by induction $M(0)$ holds which, by Lemma 1.1, is just the statement of the theorem.

Corollary 3.2: Let R be a left fully bounded Goldie, J-Noetherian ring and M an R-module with $r\text{-}\mathrm{rk}\, M \geq \text{J-dim}\, R + 1$. Suppose $M \oplus R = N \oplus R$ for some module N. Then $M \cong N$.

Proof: We include the proof for completeness. This is the proof given in [9, Corollary 2.6].

Let $\sigma: N \oplus R \longrightarrow M \oplus R$ be the given isomorphism and let $\alpha \oplus t = \sigma(0 \oplus 1)$. By Theorem 3.1 there exists an automorphism σ_1 of $M \oplus R$ such that, if $\alpha' \oplus t = \sigma_1(\alpha \oplus t)$, then α' is a unimodular element of M. Thus there exists an automorphism σ_2 of $M \oplus R$ such that $\sigma_2(\alpha' \oplus t) = \alpha' \oplus 1$ and an automorphism σ_3 of $M \oplus R$ such that $\sigma_3(\alpha' \oplus 1) = 0 \oplus 1$. So, by replacing σ by $\sigma_3\, \sigma_2\, \sigma_1\, \sigma$, we may assume that $\sigma(0 \oplus 1) = 0 \oplus 1$. But now
$N \cong N \oplus R/0 \oplus R \cong \sigma(N \oplus R)/\sigma(0 \oplus R) = M \oplus R/0 \oplus R \cong M$, as required.

Corollary 3.3: Let C be a commutative Noetherian ring and R a finite C-algebra with $\text{J-dim}\, R = n$. Let M be a finitely generated R-module with $f\text{-}\mathrm{rk}\, M \geq n + 1$ and suppose that $M \oplus R \cong N \oplus R$ for some module N. Then $M \cong N$.

Proof: By [5, Theorem 3.20 and Proposition 2.1], R is a fully bounded Noetherian ring. Thus this result follows from Corollary 3.2 and Lemma 1.2.

Remark 3.4: This answers the question raised in the introduction (this question has also been asked in [3, 8B]).

Theorem 3.1 also enables us to prove the following result due to R. B. Warfield, ([10, Theorem 3.10]).

Corollary 3.5: (Stable Range Theorem) Let R be a left fully bounded Goldie, J-Noetherian ring with J-dim $R = n$. Suppose for $m \geq n + 1$ that $R = \sum_{1}^{m+1} Ra_i$ for some $a_i \in R$. Then there exist $f_i \in R$ such that

$$R = \sum_{1}^{m} R(a_i + f_i a_{m+1}).$$

Proof: In Theorem 3.1 put $M = R^{(m)}$, $\alpha = (a_1, \ldots, a_m)$ and $t = a_{m+1}$.

Acknowledgement: The author would like to thank the British Science Research Council for financial support through a NATO Posdoctoral Research Fellowship.

REFERENCES

1. H. Bass, K-theory and Stable Algebra, Publ. Math. I.H.E.S., No. 22 (1964), 5-60.

2. K. A. Brown, T. H. Lenagan, and J. T. Stafford; K-theory and Stable Structure of Noetherian Group Rings, to appear.

3. D. Eisenbud and E. G. Evans, Jr., Generating Modules Efficiently: Theorems From Algebraic K-theory, J. Algebra 27 (1973), 278-305.

4. A. W. Goldie, The Structure of Noetherian Rings, in "Lectures on Rings and Modules," Lecture Notes in Mathematics No. 246, Springer-Verlag, Berlin/New York, 1971.

5. C. Procesi, "Rings With Polynomial Identity," M. Dekker, New York, 1973.

6. R. Rentschler and P. Gabriel, Sur la Dimension des Anneaux et Ensembles Ordonnés, C. R. Acad. Sci. Paris Sér A, 265 (1967) 712-715.

7. J.-P. Serre, Modules Projectifs et Espaces Fibrés à Fibre Vectorielle, Sém. Dubreil (1957-58), No. 23.

8. J. T. Stafford, Stable Structure of Noncommutative Noetherian Rings II, J. Algebra, to appear.

9. R. G. Swan, "Algebraic K-thoery," Lecture Notes in Mathematics No. 76, Springer-Verlag, Berlin/New York, 1968.

10. R. B. Warfield, Jr., Cancellation of Modules and Groups and Stable Range of Endomorphism Rings, to appear.

STABLE GENERATION OF MODULES

R. B. Warfield, Jr.
University of Washington
Seattle, WA 98195

We say that a finitely generated module over a ring R is <u>uniquely presentable</u> by a projective module P if there is an epimorphism $P \to A$ and any two such epimorphisms are <u>right equivalent</u>. That is, if f and g are two such epimorphisms, there is an isomorphism $\phi: P \to P$ such that $f = g\phi$. (All of our modules will be right modules, and homomorphisms act on the left.) In particular, if A is a finitely generated module, we let $g(A)$ be the minimum number of generators of A, and $u(A)$ the smallest integer n such that A is uniquely presentable by R^m for all $m \geq n$. It is clear that there is no reason in general to think that $u(A)$ will be finite, and our purpose will be to obtain some theorems saying that it is, and giving bounds on its size. If A is a finitely generated right R-module, we say that A is <u>stably generated</u> by n elements if given any set of generators $\{x_1, \ldots, x_t\}$ $(t>n)$, there are elements $y_i \in (x_{n+1}R + \ldots + x_tR)$ $(1 \leq i \leq n)$ such that the elements $x_i + y_i$ $(1 \leq i \leq n)$ generate A. We let $s(A)$ be the smallest integer n such that A is stably generated by n elements, with the understanding that if no such integer exists, then $s(A) = \infty$. We give several theorems giving bounds on the size of $s(A)$, which in turn give information about $u(A)$, since, as we show, $u(A) \leq g(A) + s(A)$.

The main results stated in this paper are to be proved in detail in forthcoming papers of the author. However, the point of view here is different, and a number of the results will not appear elsewhere (in particular, Proposition 3 and Theorems 4, 7, 9, and 10.) This research was supported in part by a grant from the National Science Foundation.

1. Stability and uniqueness. If A is a finitely generated module over a ring R, and I is a two-sided ideal of R, such that $AI = 0$, then A may also be regarded as an R/I-module. It is clear that $g(A)$ and $s(A)$ mean the same thing in either context, but $u(A)$ does not. For example, if $A = Z/5Z$, then it is easy to see that if A is regarded as a $Z/5Z$-module, then $u(Z/5Z) = 1$, while as a Z-module, $u(Z/5Z) > 1$. In fact, as a Z-module, $u(Z/5Z) = 2$, but this is not quite self-evident. This is a special case of a result of Levy and Robson [8,1.7], who show that if A is an Artinian module over any ring, then $u(A) \leq 2g(A)$. It is therefore a little surprising that we can get a bound for $u(A)$ in terms of the "ring independent" invariants $g(A)$ and $s(A)$, as we do in this section.

Proposition 1. If A is a module and P and Q are projective modules, and $f: P \to A$ and $g: Q \to A$ are epimorphisms, then the three epimorphisms $(f,0)$, $(0,g)$, and (f,g) from $P \oplus Q$ to A are right equivalent.

This is essentially a step in the proof of Schanuel's lemma. We omit the proof, (cf. [13,Lemma 1]).

Proposition 2. If R is a ring and A a finitely generated R-module with $s(A) = s < \infty$, and $f\colon R^{s+k} \to A$ is an epimorphism, with $k \geq 0$, then there is a change of basis in R^{s+k} such that with respect to this new basis, $f = (g,0)\colon R^s \oplus R^k \to A$. (That is, there is a basis $\{z_1,\ldots,z_{s+k}\}$ such that $f(z_i) = 0$, $s < i \leq s+k$.)

Proof. If our original basis is $\{x_1,\ldots,x_{s+k}\}$, then according to the definition of $s(A)$, there are elements $y_i \in x_{s+1}R + \ldots + x_{s+k}R$ $(1 \leq i \leq s)$ such that the elements $f(x_i+y_i)$ generate A. To start our new basis, choose $z_i = x_i + y_i$, $1 \leq i \leq s$. We now have a basis $\{z_1,\ldots z_s, x_{s+1},\ldots,x_{s+k}\}$ for our free module. For each $i,(s < i \leq s+k)$, there is an element $w_i \in z_1R+\ldots+z_sR$ with $f(w_i) = f(x_i)$. We let $z_i = x_i - w_i$, $(s < i \leq s+k)$, to complete the desired new basis.

Proposition 3. If R is a ring and A a finitely generated R-module, and $s(A) < \infty$, then $u(A) \leq g(A) + s(A)$.

Proof. Let $s = s(A)$ and $f\colon R^{s+k} \to A$ be an epimorphism, where $k \geq g(A)$. According to proposition 2, f is right equivalent to an epimorphism of the form $(g,0)\colon R^s \oplus R^k \to A$. Since $k \geq g(A)$, there is an epimorphism $(0,h)\colon R^s \oplus R^k \to A$. Proposition 1 implies that the epimorphisms $(0,h)$ and $(g,0)$ are right equivalent. Since h was chosen independently of f, the result is proved.

2. Generalizations of the theorems of Bass and Forster-Swan. The notion of the stable number of generators of a module arose first in the special case in which we regard the ring as a module over itself. In [1], Bass investigated what is in our notation $s(R_R)$. Modifying his terminology, it is now customary to say that

n is <u>in the stable range</u> for a ring R if $s(R_R) \le n$. Bass showed that a cyclic module over an Artinian ring is stably generated by one element. Suppose that R is a finite algebra over a commutative ring S. (That is, R is an S-algebra and is finitely generated as an S-module.) For simplicity, we assume that S is Noetherian, of classical Krull dimension d. Then a special case of Bass' results is that

$$s(R_R) \le d+1.$$

In an apparently different direction, we can study the number of generators of an R-module in terms of local data. For every maximal ideal M of the commutative ring S, and every finitely generated R-module A (where here, again, R is a finite S-algebra) we let

$$g(A,M) = g(A/AM).$$

A special case of the Forster-Swan theorem [11] is

$$g(A) \le d + \max\{g(A,M)\},$$

where d is again the classical Krull dimension of the commutative Noetherian ring S, (d is assumed to be finite), and the maximum is taken over all maximal ideals M of S. In [2], Eisenbud and Evans simultaneously generalized these results by showing [2,Theorem B] that in the above situation

$$s(A) \le d + \max\{g(A,M)\}.$$

There are many well known situations where these estimates are quite good. For example: If A is a finitely generated torsion-free module over a Dedekind domain, then the numbers g(A,M) are all the same, and equal to the rank of A. The minimal number of generators is either rank(A) or rank (A)+1. For modules over a principal ideal domain, the Forster-Swan estimate on the number of generators is off by one, but as an estimate on the stable number of generators, it is again sharp. (For example, Z, as a group, is not stably generated by one element. It is generated by 2 and 5, but not by any integer of the form $2 + 5n$. The theorem predicts that it is stably generated by 2 elements. Showing this by elementary number theoretic arguments is an amusing exercise.)

We will now state a version of the above result which gets away from algebras over commutative rings. We recall that a prime ring R is *right bounded* if every essential right ideal of R contains a nonzero two sided ideal. A ring is *right fully bounded* if for every prime ideal P, R/P is right bounded. The *classical Krull dimension* of a ring is computed just as in the commutative case by looking at chains of prime ideals. If R is a right Noetherian, right fully bounded ring, and M is a maximal ideal of R, then R/M is an Artinian ring, which is a useful fact to bear in mind in what follows. If A is a finitely generated R-module and M a maximal ideal such that R/M is Artinian, we define $g(A,M) = g(A/AM)$ as before. Our generalization of the Forster-Swan and Eisenbud-Evans theorems is then the following:

Theorem 1. If R is a Noetherian right fully bounded ring of finite classical Krull dimension n, and A is a finitely generated right R-module, then

$$s(A) \leq n + \max \{g(A,M)\}$$

where the maximum is taken over all maximal ideals M of R.

The detailed proof of this will appear in a forthcoming paper of the author. We will make a few remarks here about what is involved. To start with, the arguments used to prove the corresponding result for algebras over a commutative ring frequently become meaningless in a noncommutative setting, especially if localization is involved. It is therefore necessary to do the entire thing over, but when one does, one finds that one has results which are in some respects better than before.

As in the original setting, it is desirable to look at primes other than maximal ideals. For this it is necessary to know what to look at when Swan and Eisenbud-Evans look at the number of generators of a localization A_p, since in general we can't localize at a prime ideal. The solution is provided by Goldie's theorem which says that we can localize at a prime ideal in a Noetherian ring if that prime happens to be zero. Since in the commutative case the number of generators of A_p is the same as the number of generators of $(A/AP)_p$ (by Nakayama's lemma), this is actually quite reasonable. We therefore define g(P,A), the number of generators of a module at a prime P in R, to be the number of generators of

$$(A/AP) \otimes Q(R/P)$$

where Q(R/P) is the Goldie right quotient ring of R/P and this module is regarded as a Q(R/P)-module.

We can now introduce the other ideas that appear in the papers of Swan [11] and Eisenbud-Evans [2]. We will restrict ourselves to right Noetherian rings here, since a number of additional hypotheses with no parallel in the commutative case are needed to avoid it (see [14] for details). We say an ideal is a J-ideal if it is the intersection of maximal ideals, and the J-dimension of R written J-dim(R)) is computed just as the classical Krull dimension, but using only prime J-ideals. We again need to require that for all prime J-ideals P, R/P is right bounded, which implies (among other things) that if M is a maximal ideal, then R/M is Artinian. For any finitely generated right R-module A, we then define

$$b(A,P) = g(A,P) + \text{J-dim}(R/P).$$

The following result is then the analogue of the results of Swan [11] and Theorem B of Eisenbud-Evans [2], except that we retain the Noetherian hypothesis here to avoid further complications.

Theorem 2. Let R be a right Noetherian ring, such that R/P is right bounded for each prime J-ideal P, and A a finitely generated right R-module. Then

$$s(A) \leq \max \{b(P,A)\}$$

where the maximum is taken over all J-primes P.

The reader will notice that in Theorem 2 we require our ring to be right Noetherian, while in Theorem 1 it is required to be Noetherian--that is, left and right Noetherian. This is not an accident, and arises from the fact that while in the commutative situation, Theorem 1 immediately follows from Theorem 2, this is not quite so obvious in the noncommutative situation, and we can prove it only with an additional hypothesis on the other side of the ring.

We now return to one other aspect of the proofs of the theorems of Swan and Eisenbud-Evans, to indicate where the commutative methods do not apply and new methods are needed. The work of Swan and Eisenbud-Evans made essential use of a result of Bass already mentioned. Bass showed that if A is a cydic module over an Artinian semisimple ring R, and $A = xR + B$, where B is a submodule, then for some $y \in B$, $A = (x+y)R$. Swan modified this to show that you can choose y such that $A = (x + yz)R$ for any z in the center of R, $z \neq 0$. The point of this result is that we wish to modify some element x in A to make the result generate as much as possible of A_p (where P is a prime in the underlying commutative ring) without upsetting anything we may already have done at a finite number of prime ideals larger than P. The central multiplier enables us to get inside any of these previously worked-on primes and to get an element of A_p which comes from A. None of this works in our setting, and it was necessary to prove directly that a right Goldie semiprime ring has "essentially" one in the stable range. The precise result is that if R is right Goldie and semiprime and R_R has $xR + B$ as an essential right ideal, then for some $y \in B$, $(x+y)R$

is an essential right ideal of R. This was proved by the author and by J. T. Stafford [9]. It is precisely what is needed in place of the Bass-Swan argument to prove Theorem 1 and related theorems. In fact, a more general result turns out to be useful in many contexts, and we state it here.

Definition. If R is a ring, A a right R-module, F a finitely generated projective R-module, P a prime ideal of R such that R/P is right Goldie, and $f \in \mathrm{Hom}(F,A)$, then we say that f is maximal at P if the induced homomorphism

$$f^*: (F/FP) \otimes Q(R/P) \to (A/AP) \otimes Q(R/P)$$

is either injective or surjective.

This means that f* has maximal rank at those primes P. Note that if F = R, and if $f: R \to A$ is maximal at P, then certainly f(1) is basic at P in the sense of Eisenbud and Evans, and our requirement is actually stronger than that.

Theorem 3. ([14, Theorem 1]. Let R be a ring, X a finite set of primes of R such that $P \in X$ implies that R/P is right Goldie, F a finitely generated projective module, A and G modules, and $f \in \mathrm{Hom}(F,A)$ and $g \in \mathrm{Hom}(G,A)$. Then there is a homomorphism $\phi: F \to G$ such that for all primes $P \in X$, either $f + g\phi: F \to A$ is maximal at P, or it has the maximal possible rank--that is, the image of the induced map $(f + g\phi)_p:(F/FP) \otimes Q \to (A/AP) \otimes Q$ (where Q is the right quotient ring of R/P) contains the image of $(f,g)_P: (F/FP \oplus G/GP) \otimes Q \to (A/AP) \otimes Q$.

Using this lemma, we obtain in [14] an improved version of Theorem 2, in which we obtain not only results on epimorphisms $R^n \to A$ but also results on epimorphisms $F^n \to A$ for any desired finitely generated projective module F.

It would be nice to remove the boundedness hypotheses on Theorems 1 and 2. In [14] we give an analogue of Swan's theorem, for prime rings of Krull dimension one without the boundedness hypotheses. Presumably one should be able to prove a corresponding theorem for Noetherian rings satisfying Stafford's condition of ideal invariance [9]. However, for prime rings of Krull dimension one, we have not been able to prove the analogue of Theorem 1, though we can prove $s(A) \leq 2 + \max \{g(A,M)\}$.

In a different direction, we can say that a finitely generated module A is stably generated by a projective module P if for every epimorphism $(f,h): P \oplus Q \to A$ there is a homomorphism $\alpha: P \to Q$ such that $(f+h\alpha): P \to A$ is an epimorphism. One should be able to prove that a "large enough" projective has this property. In effect, our methods require that we assume in addition a direct decomposition of P into small pieces. In the special case in which $A = R_R$, the theorem we suggest should yield a proof of Serre's theorem on free summands of projectives, and of Stafford's generalizations of this result to noncommutative rings [10].

3. What does the stable range of a ring say about its modules? There are many rings for which results like those in the previous section have not been proved, but for which one does know something about the stable range of the ring. Examples are Heitman's theorem [5] that any commutative ring of classical Krull dimension d has $d+2$ in the stable range, and Stafford's theorem [9] that a Noetherian ring which is ideal invariant and of (noncommutative) Krull dimension d has $d+1$ in the stable range. For such rings, one can still say something about $s(A)$, for finitely generated modules A, though the results are slightly weaker than those in the previous section. For example, in Theorem 1, by throwing

away part of the information, we see that if A is a finitely generated module over a Noetherian right fully bounded ring of Krull dimension d, then $s(A) \leq g(A) + d$. This is the sort of estimate that turns out to be valid more generally. Bearing in mind that such a ring has $d+1$ in the stable range, we see that the weak form of Theorem 1 is a special case of the next theorem.

Theorem 4. Let R be a ring with n in the stable range and A a finitely generated right R-module. Then

$$s(A) \leq g(A) + n - 1.$$

Proof. Since if there is an epimorphism $A \to B$, then $s(B) \leq s(A)$, it will suffice to prove Theorem 4 in the special case in which A is free, say $A = R^k$.

Let $(f,h)\colon R^{k+m} \oplus H \to R^k$ be an epimorphism, where $m \geq n-1$ and we may assume that $k > 1$, since the case in which $k = 1$ is what is given as our hypothesis. We will proceed by induction. Let $\pi\colon R^k \to R$ be a projection with kernel A, $A \cong R^{k-1}$. Since R has n in the stable range, there is a homomorphism $\alpha\colon R^{k+m} \to H$ with $\pi(f + h\alpha)\colon R^{k+m} \to R$ an epimorphism. Let $\sigma\colon R \to R^{k+m}$ be a splitting. We then have $R^{k+m} = \sigma(R) \oplus B$, and the cancellation theorem of [12,1.2] implies that $B \cong R^{k+m-1}$. Here, $B = \mathrm{Ker}[\pi(f+h\alpha)]$. We let $A' = (f+h\alpha)\,(\sigma(R))$, and note that

$$R^k = A \oplus A'.$$

Let δ be the projection of R^k onto A with kernel A'. An easy computation shows that $(f+h\alpha,g)\colon R^{k+m} \oplus H \to R^k$ is again an epimorphism, and since this epimorphism takes $\sigma(R)$ onto A', the map $\delta(f+h\alpha,g)$ may be regarded as giving an epimorphism

$$B \oplus H \to A.$$

Since we know that $B \cong R^{k+m-1}$ and $A \cong R^{k-1}$, we know by induction that there is a homomorphism $\beta\colon B \to H$ such that if ϕ is the restriction of $f + h\alpha$ to B, then $\delta(\phi + h\beta)(B) = A$. We extend β to a homomorphism $\beta' : R^{k+m} \to H$ by letting it equal β on B and 0 on $\sigma(R)$. We then have a homomorphism

$$f + h(\alpha + \beta')\colon R^{k+m} \to R^k$$

which a computation shows to be an epimorphism, as required. (In detail, $[f+h(\alpha+\beta')](\sigma(R)) = A'$, so A' is in the image of the alleged epimorphism. It therefore suffices to show that the induced map onto R^k/A' is an epimorphism, i.e., that $\delta[f+h(\alpha+\beta')](R^{k+m}) = A$. This is true since by our construction we already know that $\delta[f+h(\alpha+\beta')](B) = A$.)

Using the results of section 1, the reader will see that one also obtains an estimate for $u(A)$, namely

$$u(A) \le 2g(A) + n - 1.$$

This estimate, however, can be improved by using different methods, as we see in the next section.

4. Redundancy in presentations, and another uniqueness result. If A is a finitely generated module and A is stably generated by $s = s(A)$ elements, then for any epimorphism $f\colon R^m \to A$ with $m > s(A)$, there is a basis change in R^m so that $R^m = R^s \oplus R^{m-s}$ and such that the epimorphism f restricts to zero on the second factor. This is what we have in mind when we say that a presentation involving more than $s(A)$ generators is redundant. From the previous section, it follows that if R is a ring with n in the stable range, then any presentation for A involving more than $g(A) + n - 1$ generators displays this redundancy, and that any such presentation can be reduced to a presentation by $g(A) + n - 1$ elements by a basis change. It turns out that if we are only interested in this redundancy, we can improve this result by using

different methods, and also get a lower estimate for $u(A)$ which does not use $s(A)$.

To state our results, it is convenient to work in the category $\underline{L}$ whose objects are triples (A,B,f), where A and B are right R-modules and $f\colon A \to B$ is a homomorphism. A morphism in this category between objects (A,B,f) and (A',B',f') is given by a pair of homomorphisms, $\alpha\colon A \to A'$ and $\beta\colon B \to B'$ such that $\beta f = f'\alpha$. If, then, we fix A and B and consider two objects in our category, (A,B,f) and (A,B,g) then they are isomorphic if and only if there are automorphisms α and β of A and B such that $g = \beta f \alpha^{-1}$. This corresponds to the usual notion of equivalence of matrices. However, there is another notion of equivalence which is stronger and requires that $\beta = 1$. This appears, for example, in the notion of a projective cover. As in section 1, we say that (A,B,f) and (A',B,f') are <u>right equivalent</u> if there is an isomorphism (α,β) between them such that $\beta = 1$. We regard it as obvious that $\underline{L}$ is an additive category. In particular, if (A,B,f) and (A',B',f') are objects, then their direct sum in $\underline{L}$ is the triple $(A\oplus A', B\oplus B', \begin{pmatrix} f & 0 \\ 0 & f' \end{pmatrix})$. In some special cases, this appears differently. If we add $(R,0,0)$ to a triple of the form (R^n,R^m,f), it has the effect of adding a column of zeros to the matrix and leaving the number of rows alone. Proposition 1 says in the language of $\underline{L}$ that if $f\colon R^k \to A$ and $g\colon R^m \to A$ are epimorphisms, then the epimorphisms

$$(f,0) : R^k \oplus R^m \to A$$

and

$$(0,g) : R^k \oplus R^m \to A$$

are right equivalent. We have therefore, in particular, two objects in an additive category which become isomorphic once one has added to each a large number of copies of a particular object--in this case the object $(R,0,0)$. The question that arises is this: is there a cancellation theorem which can be used to remove redundant copies of the object $(R,0,0)$? Further, we would like to show that if m is sufficiently larger than k, then the epimorphism $g: R^m \to A$ is right equivalent to $(f,0): R^k \oplus R^{m-k} \to A$.

Both of these projects can be carried out if there is some integer n which is in the stable range for R. The basic cancellation result, proved in [12], is that if S, Y, and A are objects in an additive category, if $X \oplus A^m = Y \oplus A^{m+n}$, where n is an integer in the stable range of the endomorphism ring of A, then $X \cong Y \oplus A^n$. Since the endomorphism ring of the object $(R,0,0)$ in $\underline{L}$ is the ring R itself, this is most of what is needed to prove the following result, which is closely related to Theorems 8 and 9 in [13].

Theorem 5. If R is a ring having n in the stable range and A a finitely generated R-module, and $f: R^m \to A$ and $h: R^k \to A$ are epimorphisms, where $m \geq n + k$, then f is right equivalent to the epimorphism $(h,0): R^k \oplus R^{m-k} \to A$.

This theorem is not quite proved by the considerations preceeding it, because a cancellation theorem would appear to tell us less than we wanted. It would seem that what we can conclude is only that the objects (R^m,A,f) and $(R^k \oplus R^{m-k},A,(h,0))$ are isomorphic in the category $\underline{L}$. This would involve an automorphism of A, and for right equivalence, we require this automorphism to be the identity. The point is that in [12], where the connection between cancellation and the stable range of endomorphism rings is first made, what is proved is not just a cancellation result, but a strong form of cancellation known as the n-substitution property. This is a little complicated, and we refer to [12] and [13] for details in general. (We give a more detailed treatment of the case $n = 1$ below.) Suffice it to say here that when applied to the category $\underline{L}$, this is precisely what is needed to guarantee that the isomorphism we end up with between (R^m,A,f) and $(R^k \oplus R^{m-k},A,(h,0))$ is actually a right equivalence, as required.

Theorem 6. If A is a finitely generated module over a ring R with n in the stable range, then

$$u(A) \leq g(A) + n$$

Proof. If $m = k + n$, and $k \geq g(A)$, and we pick a fixed epimorphism $h: R^k \to A$, then Theorem 5 shows that every epimorphism $f: R^{k+n} \to A$ is right equivalent to $(h,0): R^k \oplus R^n \to A$. Hence all such epimorphisms are right equivanent.

In certain circumstances the estimate in Theorem 5 can be improved. In particular, if a ring R has one in the stable range, then any object in an additive category with endomorphism ring isomorphic to R actually can be cancelled. We will elaborate on this point slightly. We say an object A has the substitution (or 1-substitution) property if given an object M with two decompositions

$$M = A_1 \oplus X = A_2 \oplus Y.$$

with $A \cong A_1 \cong A_2$, there is a subobject C such that

$$M = C \oplus X = C \oplus Y.$$

Clearly this implies that A can be cancelled from direct sums, but it is stronger than that. In [12] it is proved that this holds for an object A in an additive category if and only if the ring End(A) has 1 in the stable range. The reader may easily verify that if we are working in the category $\underline{L}$ and A is the object $(R,0,0)$ then this substitution property implies that the objects X and Y are right equivalent (and not just isomorphic) objects in $\underline{L}$. We therefore have the following special case, which is actually slightly better than what we get by substituting $d = 1$ in Theorem 5.

Theorem 7. If R is a ring with 1 in the stable range and $f: R^m \to A$ and $g: R^m \to A$ are epimorphisms, then f and g are right equivalent. Further, $g(A) = s(A) = u(A)$.

All of this follows from the above remarks, except the statement about $s(A)$, which follows from Theorem 4. It follows from our earlier comments that Artinian rings, and rings which are Artinian modulo their Jacobson radical have 1 in the stable range. A Von Neumann regular ring has 1 in the stable range if and only if it is unit regular [6], and, in particular, commutative von Neumann regular rings have this property. More generally [4], a finite algebra over a commutative ring S such that S modulo its Jacobson radical is von Neumann regular has 1 in the stable range.

5. Matrices over Bezout rings. Methods similar to those in the previous section are used in [13] to study the equivalence of matrices over rings with some integer in the stable range. In particular, if two matrices are given of the same size, one can ask how much the matrices have to be enlarged by the addition of suitable zeros and ones to get equivalent matrices. This enquiry was initiated by Fitting in [3], and we refer to [13] for a variety of results in this direction. In this section we will confine ourselves to remarks in the special case in which the ring has 1 in the stable range. The first result is a special case of those in [13], but the applications are new.

Theorem 8. If R is a ring with 1 in the stable range and A and B are matrices of the same size and with isomorphic cokernels, then A and B are equivalent.

Proof. If f and g are the corresponding homomorphisms, $R^n \to R^m$, then we have two exact sequences which fit into the following diagram:

$$\begin{array}{ccccccc} R^n & \xrightarrow{f} & R^m & \to & M & \to & 0 \\ & & & & \| & & \\ R^n & \xrightarrow{g} & R^m & \to & M & \to & 0 \end{array}$$

Theorem 7 implies that there is an isomorphism $\alpha: R^m \to R^m$ giving a right equivalence of the two epimorphisms $R^m \to M$. It follows, in particular, that α identifies Image(f) with image(g), and hence Theorem 7 again applies, giving an isomorphism $\beta: R^n \to R^n$ such that $\alpha f = g\beta$. This proves the result.

We give two applications to Bezout rings, one using right equivalence (and therefore Theorem 7) and the other equivalence of matrices (an application of Theorem 8). We recall that a ring is a right Bezout ring if every finitely generated right ideal is principal, and a right Hermite ring if every matrix over R is right equivalent to a lower triangular matrix. (Equivalently, R is right Hermite if for every matrix A there is an invertible matrix U such that AU is lower triangular.) In [7], Kaplansky points out that it is enough to do this for a 2×1 matrix A. Kaplansky proves a number of theorems about Hermite rings, including the theorem that a left and right Bezout domain is left and right Hermite [7,3.4].

Theorem 9. A right Bezout ring with 1 in the stable range is right Hermite. In particular, this applies to a right Bezout ring which is semilocal, or, more generally, one which modulo its Jacobson radical is unit regular.

Proof. Let (a,b) be a 2×1 matrix representing a homomorphism $R^2 \to R$, and let the image of this homomorphism be cR. The matrix $(c,0)\colon R^2 \to R$ has the same image. Theorem 7 implies that there is an isomorphism $\alpha: R^2 \to R^2$ such that $(a,b)\,\alpha = (c,0)$. This proves that (a,b) is right equivalent to a lower triangular matrix, and proves that R is right Hermite.

Theorem 10. Let R be a right Bezout ring with 1 in the stable range and suppose that every finitely presented right R-module A is a direct sum of g(A) cyclic modules. Then any square matrix over R is equivalent to a diagonal matrix.

Proof. If $f: R^n \to R^n$ is a homomorphism, and coker(F) = A, then $n \geq g(A)$, and there is a diagonal matrix $h: R^g \to R^g$ such that $g = g(A)$ and coker(h) $\cong$ A. Expanding h by adding on a suitable unit matrix, we obtain a diagonal $n \times n$ matrix with cokernel isomorphic to A. A reference to Theorem 8 completes the proof.

We do not wish to spend time exploring the hypotheses of this theorem. We remark that any von Neumann regular ring is an example of a right Bezout ring in which every finitely presented module A is a direct sum of g(A) cyclic modules. Examples of von Neumann regular rings R such that $R_R \cong R_R \oplus R_R$ show that not all von Neumann regular rings are right Hermite. It is a remark of Kaplansky's that a von Neumann regular ring has 1 in the stable range if and only if it is unit regular (that is, for any x there is a unit u with $xux = x$). This appears in [6], in which Henriksen proves Theorems 9 and 10 for unit regular rings.

REFERENCES

[1] H. Bass, "K-theory and stable algebra," Pub. Math. I. H. E. S. 22 (1964), 5-60.

[2] D. Eisenbud and E. G. Evans, Jr., "Generating modules efficiently: theorems from algebraic K-theory," J. Alg. 27 (1973) 278-305.

[3] H. Fitting, "Uber den Zusammenhang zwischen dem Begriff der Gleichartigkeit zweier Ideale und dem Aquivalenzbegriff der Elementarteilertheorie," Math. Ann. 112 (1936), 572-582.

[4] K. R. Goodearl and R. B. Warfield, Jr., "Algebras over zero-dimensional rings," Math. Ann. 223 (1976), 157-168.

[5] R. Heitman, "Generating ideals in Prufer domains," Pac. J. Math. 62 (1976), 117-126.

[6] M. Henriksen, "On a class of regular rings which are elementary divisor rings," Archiv der Math. 24 (1973) 133-141.

[7] I. Kaplansky, "Elementary divisors and modules," Trans. Amer. Math. Soc. 66 (1949), 464-491.

[8] L. Levy and J. C. Robson, "Matrices and pairs of modules," J. Alg. 29 (1974), 103-121.

[9] J. T. Stafford, "Stable structure of non-commutative Noetherian rings," J. Alg. 47 (1977), 244-267.

[10] J. T. Stafford, "Stable structure of non-commutative Noetherian rings, II," J. Alg. (to appear).

[11] R. Swan, "The number of generators of a module," Math. Zeit. 102 (1967), 318-322.

[12] R. B. Warfield, Jr., "Cancellation of modules and groups and stable range of endomorphism rings," (to appear).

[13] R. B. Warfield, Jr., "Stable Equivalence of Matrices and Resolutions," Comm. Alg. (to appear, 1978).

SOME ASPECTS OF FULLER'S THEOREM

Goro Azumaya
Indiana University
Bloomington, Indiana 47401

Let R, S be rings with unit element. By R- or S-modules we shall always mean unital modules. Let $U = {}_RU_S$ be an R-S-bimodule. Then for every left R-module X a canonical homomorphism

$$\rho(X): {}_RU \otimes_S \mathrm{Hom}_R(U, X) \longrightarrow {}_RX$$

is defined by $\rho(X)(u \otimes f) = f(u)$ for $u \in U$, $f \in \mathrm{Hom}_R(U, X)$. Similarly, for every left S-module Y a canonical homomorphism

$$\sigma(Y): {}_SY \longrightarrow {}_S\mathrm{Hom}_R(U, U \otimes_S Y)$$

is defined by $(\sigma(Y)y)u = u \otimes y$ for $y \in Y$, $u \in U$. The natural transformations ρ and σ are fundamental tools in the categorical theory of modules. Indeed, the theory of Morita equivalence is precisely for the case where both ρ and σ are isomorphisms. Generalizing the Morita theory, Fuller [1] considered the case where $\sigma(Y)$ is an isomorphism for all left S-modules Y and $\rho(X)$ is an isomorphism for all X in a certain class of left R-modules, and succeeded in obtaining a theorem characterizing the structure of U which corresponds to this case. On the other hand, Sato [5] has recently worked out determining the type of U for which $\sigma(Y)$ is an isomorphism for all left S-modules Y, and as an application given an improvement and sharpening of Fuller's theorem. In the present note, by observing ρ rather than σ, we attempt to get another approach, which, combined with Sato's results, yields a further refinement and clarification of Fuller's characterization.

Let $\overline{X}$ denote, for each left R-module X, the image of $\rho(X)$ i.e. the sum of all homomorphic images of ${}_RU$ in X. Then clearly $Hom_R(U, \overline{X}) = Hom_R(U, X)$, and this implies that $\rho(\overline{X})$ is an isomorphism if and only if $\rho(X)$ is a monomorphism. Let $Gen_R(U)$ be the class of those left R-modules X for which $\overline{X} = X$. It follows then that <u>$\rho(X)$ is an isomorphism for all X in $Gen_R(U)$ if and only if $\rho(X)$ is a monomorphism for all left R-modules X</u>. Now that X is in $Gen_R(U)$ means that X is a sum of homomorphic images of ${}_RU$, and this is also equivalent to the condition that X is a homomorphic image of a direct sum of copies of ${}_RU$, that is, there exist an index set A and an epimorphism ${}_RU^{(A)} \to {}_RX$, where $U^{(A)}$ means the A-times direct sum of U. Generally, each homomorphism $h: {}_RU^{(A)} \to {}_RX$ can be identified with a family $\{h_\alpha \mid \alpha \in A\}$ of homomorphisms $h_\alpha: {}_RU \to {}_RX$ such that $h(\{u_\alpha\}) = \sum h_\alpha(u_\alpha)$ for every $\{u_\alpha\} \in U^{(A)}$. (Here, $u_\alpha = 0$ for all but a finite number of α, while h_α's need not satisfy such a condition.)

<u>Lemma 1</u>. Let $h = \{h_\alpha\}$ be an epimorphism ${}_RU^{(A)} \to {}_RX$ and let $\rho(X)$ be a monomorphism. Then $U \otimes_S Hom_R(U, X) = U \otimes \sum Sh_\alpha$.

<u>Proof</u>. Let t be any element of $U \otimes_S Hom_R(U, X)$ and let $x \in X$ be the image of t by $\rho(X)$. Since $h: U^{(A)} \to X$ is an epimorphism, there exists $\{u_\alpha\} \in U^{(A)}$ such that $x = h(\{u_\alpha\}) = \sum h_\alpha(u_\alpha)$. Consider now $\sum u_\alpha \otimes h_\alpha \in U \otimes_S Hom_R(U, X)$. Its image by $\rho(X)$ is also $\sum h_\alpha(u_\alpha) = x$. Since however $\rho(X)$ is a monomorphism, it follows $t = \sum u_\alpha \otimes h_\alpha$, which shows that $U \otimes_S Hom_R(U, X) = \sum U \otimes h_\alpha = U \otimes \sum Sh_\alpha$.

<u>Proposition 2</u>. Let U_S be flat, and let $\rho(X)$ be a monomorphism for all left R-modules X. Then $\mathrm{Gen}_R(U)$ is closed under submodules.

<u>Proof</u>. Let A be any set and let K be a submodule of ${}_RU^{(A)}$. It suffices to show that $K \in \mathrm{Gen}_R(U)$. Put now $X = U^{(A)}/K$. Then we have an exact sequence

$$0 \longrightarrow {}_RK \xrightarrow{i} {}_RU^{(A)} \xrightarrow{h} {}_RX \longrightarrow 0,$$

where $h = \{h_\alpha\}$ is the natural epimorphism and i is the inclusion map. The submodule $\sum' Sh_\alpha$ of ${}_S\mathrm{Hom}_R(U, X)$ is a homomorphic image of the free left S-module $S^{(A)}$ by the epimorphism φ defined by $\varphi(\{s_\alpha\}) = \sum' s_\alpha h_\alpha$. Similarly, there exist a free left S-module $S^{(B)}$ and an epimorphism $\psi : {}_SS^{(B)} \longrightarrow {}_S\mathrm{Ker}(\varphi)$. Since the sequence $S^{(B)} \xrightarrow{\psi} S^{(A)} \xrightarrow{\varphi} \sum' Sh_\alpha \longrightarrow 0$ is exact and the functor $U \otimes_S$ is right exact, we have the following exact sequence:

$$U \otimes_S S^{(B)} \xrightarrow{U\otimes\psi} U \otimes_S S^{(A)} \xrightarrow{U\otimes\varphi} U \otimes_S \textstyle\sum' Sh_\alpha \longrightarrow 0.$$

If we denote by j the inclusion map $\sum' Sh_\alpha \longrightarrow \mathrm{Hom}_R(U, X)$, then, since U_S is flat, $U\otimes j : U \otimes_S \sum' Sh_\alpha \longrightarrow U \otimes_S \mathrm{Hom}_R(U, X)$ is a monomorphism and indeed an isomorphism because of Lemma 1. Combining this with another isomorphism $\rho(X)$, we have an isomorphism $p = \rho(X) \circ (U\otimes j) : {}_R(U \otimes_S \sum' Sh_\alpha) \longrightarrow {}_RX$. Consider now the following diagram:

$$\begin{array}{ccccccc}
U \otimes_S S^{(B)} & \xrightarrow{U\otimes\psi} & U \otimes_S S^{(A)} & \xrightarrow{U\otimes\varphi} & U \otimes_S \sum Sh_\alpha & \longrightarrow & 0 \\
\downarrow{\scriptstyle q_B} & & \downarrow{\scriptstyle q_A} & & \downarrow{\scriptstyle p} & & \\
U^{(B)} & \dashrightarrow{g} & U^{(A)} & \xrightarrow{\quad h \quad} & X & \longrightarrow & 0,
\end{array}$$

where q_B and q_A are canonical isomorphisms. Its right-half is commutative, because $p \circ (U\otimes\varphi)(U\otimes\{s_\alpha\}) = p(u \otimes \sum' s_\alpha h_\alpha) = \sum' h_\alpha(us_\alpha) = h(\{us_\alpha\}) = h \circ q_A(u \otimes \{s_\alpha\})$ for $u \in U$, $\{s_\alpha\} \in S^{(A)}$. On the other hand, it is possible, since both q_A and q_B are isomorphisms, to find a homomorphism $g: {}_RU^{(B)} \to {}_RU^{(A)}$ making the left-half of the (whence the whole) diagram commutative. Since the upper row is exact, it follows that the lower row is also exact, that is, the image of $U^{(B)}$ by g coincides with K, the kernel of h. Thus the proof is completed.

<u>Lemma 3</u>. Let ${}_RU$ be quasi-projective and let $S = \mathrm{End}_R(U)$. Then for every finitely generated left ideal L of S we have ${}_SL \cong {}_S\mathrm{Hom}_R(U, UL)$ canonically.

<u>Proof</u>. Let ${}_SL$ be generated by $a_1, a_2, \ldots, a_n$: $L = \sum Sa_i$. Then $UL = \sum Ua_i$. If we denote by U^n the n-times direct sum of U, then we can define an epimorphism $h: {}_RU^n \to {}_RUL$ by $h(u_1, \ldots, u_n) = \sum u_i a_i$ for $(u_1, \ldots, u_n) \in U^n$. Let $f \in \mathrm{Hom}_R(U, UL)$. Then, since ${}_RU$ is U^n-projective by Robert[4 , Proposition 1], there exists a homomorphism $g: {}_RU \to {}_RU^n$ which makes the following diagram commutative:

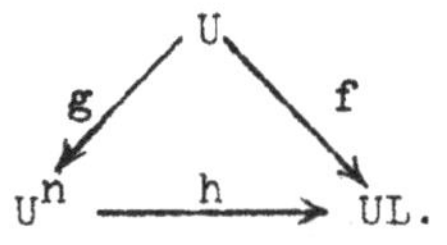

Taking the i-th entry of g for each i, we have an endomorphism $s_i \in S$ of ${}_RU$, and thus $g(u) = (us_1, \ldots, us_n)$ for $u \in U$. Therefore, we have $f(u) = h(g(u)) = h(us_1, \ldots, us_n) = \sum us_i a_i = u\sum s_i a_i$ for $u \in U$, which means that f is obtained by the right multiplication of the element $a = \sum s_i a_i$ of L. The mapping $f \mapsto a$ clearly gives a canonical isomorphism ${}_S\mathrm{Hom}_R(U, UL) \cong {}_SL$.

<u>Proposition 4</u>. Let ${}_RU$ be quasi-projective and let $S = \mathrm{End}_R(U)$. If besides $\rho(X)$ is a monomorphism for all submodules X of ${}_RU$, then U_S is flat.

<u>Proof</u>. Let L be a finitely generated left ideal of S. By Lemma 3 there is an isomorphism $\theta : {}_SL \longrightarrow {}_S\mathrm{Hom}_R(U, UL)$ which satisfies $\theta(a)u = ua$ for $a \in L$, $u \in U$. By tensoring with U_S we have then an isomorphism $U \otimes \theta : {}_RU \otimes_S L \longrightarrow {}_RU \otimes {}_S\mathrm{Hom}_R(U, UL)$. Since UL is a submodule of ${}_RU$, $\rho(UL) : {}_RU \otimes {}_S\mathrm{Hom}_R(U, UL) \longrightarrow {}_RUL$ is a monomorphism by assumption, it follows that their product $\rho(UL) \circ (U \otimes \theta) : {}_RU \otimes_S L \longrightarrow {}_RUL$ is also a monomorphism. But this is indeed the canonical epimorphism whence isomorphism, because $\rho(UL)((U \otimes \theta) \cdot (u \otimes a)) = \rho(UL) \cdot (u \otimes \theta(a)) = \theta(a)u = ua$ for $u \in U$, $a \in L$. Thus U_S is flat by Lambek [3, Proposition 1, p. 132].

Following Sato [5], U_S is called a <u>weak generator</u> if $U \otimes_S Y = 0$ for a left S-module Y implies $Y = 0$. If U_S is a weak generator then it is faithful, because we have $U \otimes_S Sa = Ua \otimes 1$ for every $a \in S$.

<u>Proposition 5</u>. Let $h = \{h_\alpha\} : {}_RU^{(A)} \longrightarrow {}_RX$ be an epimorphism and let $\rho(X)$ be a monomorphism. If besides U_S is a weak generator then $\mathrm{Hom}_R(U, X) = \sum Sh_\alpha$, that is, for each $f \in \mathrm{Hom}_R(U, X)$ there exists $\{s_\alpha\} \in S^{(A)}$ such that $f(u) = h(\{us_\alpha\})$ for all $u \in U$.

<u>Proof</u>. Let $Y = \mathrm{Hom}_R(U, X)/\sum Sh_\alpha$. Then the sequence

$$ {}_S\textstyle\sum Sh_\alpha \xrightarrow{\ j\ } {}_S\mathrm{Hom}_R(U, X) \xrightarrow{\ q\ } {}_SY \longrightarrow 0 $$

is exact, with the inclusion map j and the natural epimorphism q. Tensoring with U_S then yields an exact sequence

$$U \otimes_S \sum Sh_\alpha \xrightarrow{U\otimes j} U \otimes_S \mathrm{Hom}_R(U,\ X) \xrightarrow{U\otimes q} U \otimes_S Y \longrightarrow 0.$$

By Lemma 1 $U \otimes j$ is an epimorphism, or equivalently, $U \otimes_S Y = 0$. Since U_S is a weak generator, this implies that $Y = 0$, i.e. $\mathrm{Hom}_R(U,\ X) = \sum Sh_\alpha$.

Corollary 6. Let U_S be a weak generator and let $\rho(U^{(A)})$ be a monomorphism. Then ${}_S\mathrm{Hom}_R(U,\ U^{(A)}) \cong {}_S S^{(A)}$ canonically, and in particular $S = \mathrm{End}_R(U)$.

Proof. Apply Proposition 5 to the particular case where $X = U^{(A)}$ and $h\colon U^{(A)} \longrightarrow U^{(A)}$ is the identity map. Then for each $f \in \mathrm{Hom}_R(U,\ U^{(A)})$ there exists $\{s_\alpha\} \in S^{(A)}$ such that $f(u) = \{us_\alpha\}$ for all $u \in U$, and since U_S is a weak generator whence faithful such $\{s_\alpha\}$ is uniquely determined by f.

Corollary 7. Let U_S be a weak generator and let $\rho(X)$ be a monomorphism for all left R-modules X. Then ${}_RU$ is $\sum$-quasi-projective (i.e. $U^{(A)}$-projective for all sets A).

Proof. This is an immediate consequence of Proposition 5.

Proposition 8. In order that ${}_RU$ be finitely generated quasi-projective and $S = \mathrm{End}_R(U)$ it is necessary and sufficient that ${}_RU$ be $\sum$-quasi-projective and ${}_S\mathrm{Hom}_R(U,\ U^{(A)}) \cong {}_S S^{(A)}$ canonically for all sets A.

Proof. This is proved by generalizing or modifying Sato's technique in the proof of [5, Theorem 3.1]; observe however that we do not assume the closedness of $\mathrm{Gen}_R(U)$ under submodules. Let namely

$\{U_\alpha \mid \alpha \in A\}$ be the family of all cyclic submodules of $_RU$, and let h: $_R\sum \oplus U_\alpha \longrightarrow {}_RU$ be the epimorphism defined by $h(\{u_\alpha\}) = \sum u_\alpha$ for all $\{u_\alpha\} \in \sum \oplus U_\alpha$. Assume that $_RU$ is $U^{(A)}$-projective. Then, since $\sum \oplus U_\alpha$ is a submodule of $_RU^{(A)}$, $_RU$ is $\sum \oplus U_\alpha$-projective too (Robert [4, Proposition 1]). Hence there is a homomorphism g: $_RU \longrightarrow {}_R\sum \oplus U_\alpha$ such that $h \circ g = 1$, the identity map of U. Assume further that $_S\mathrm{Hom}_R(U, U^{(A)}) \cong {}_SS^{(A)}$ canonically. Thus, regarding g as a homomorphism $_RU \longrightarrow {}_RU^{(A)}$, there exists $\{s_\alpha\} \in S^{(A)}$ such that $g(u) = \{us_\alpha\}$ for all $u \in U$. It follows then that $Us_\alpha \subset U_\alpha$ for all $\alpha \in A$ and that for a suitable finite subset F of A $s_\alpha = 0$ whenever $\alpha \notin F$. This shows that $u = h(g(u)) = \sum_{\alpha \in A} us_\alpha = \sum_{\alpha \in F} us_\alpha \in \sum_{\alpha \in F} U_\alpha$ for all $u \in U$ and thus $_RU = \sum_{\alpha \in F} U_\alpha$ is finitely generated. Since every $\sum$-quasi-projective module is quasi-projective, this proves the sufficiency of our proposition. The necessity follows from Fuller and Hill [2, Corollary 3.2] and the well-known fact that $_S\mathrm{Hom}_R(U, U^{(A)}) \cong {}_SS^{(A)}$ canonically if $_RU$ is finitely generated.

Proposition 9. Let $\sigma(Y)$ be an epimorphism for all left S-modules Y and let $\mathrm{Gen}_R(U)$ be closed under submodules. Then $\rho(X)$ is a monomorphism for all left R-modules X.

Proof. This can also be proved in the similar manner as for the proof of (2)$\Longrightarrow$(3) in Sato [5, Theorem 2.1]. Let, for brevity, denote by H and T the functors $\mathrm{Hom}_R(U, -)$ and $U \otimes_S -$ respectively. Then for each left S-module Y there arise two natural homomorphisms $T(\sigma(Y))$: $T(Y) \longrightarrow T(H(T(Y)))$ and $\rho(T(Y))$: $T(H(T(Y))) \longrightarrow T(Y)$. It is easy to see that $\rho(T(Y)) \circ T(\sigma(Y)) = 1$, the identity map of $T(Y)$, so in particular $T(\sigma(Y))$ is a monomorphism and $\rho(T(Y))$ is an epimorphism. But since $\sigma(Y)$ is an epimorphism by assumption and since T is a right exact functor, $T(\sigma(Y))$ is also an epimorphism whence an isomorphism. It follows from this that $\rho(T(Y))$ is the inverse map of $T(\sigma(Y))$ and hence an isomorphism too. Thus we have that $\rho(X)$ is an isomorphism whenever ${}_RX \cong {}_RT(Y) = {}_RU \otimes_S Y$ for some left S-module Y.

Let now X be in $\mathrm{Gen}_R(U)$. Then there exist an index set A and an epimorphism $h: {}_RU^{(A)} \longrightarrow {}_RX$. Put $K = \mathrm{Ker}(h)$. Then K is in $\mathrm{Gen}_R(U)$ by assumption, and so there exist also an index set B and an epimorphism $g: {}_RU^{(B)} \longrightarrow {}_RK$. We then consider the following commutative diagram of left R-modules:

$$\begin{array}{ccccccc} T(H(U^{(B)})) & \xrightarrow{T(H(g))} & T(H(U^{(A)})) & \xrightarrow{T(H(h))} & T(\mathrm{Coker}(H(g))) & \longrightarrow & 0 \\ \downarrow{\scriptstyle \rho(U^{(B)})} & & \downarrow{\scriptstyle \rho(U^{(A)})} & & & & \\ U^{(B)} & \xrightarrow{\quad g \quad} & U^{(A)} & \xrightarrow{\quad h \quad} & X & \longrightarrow & 0. \end{array}$$

Its rows are exact, since T is right exact, while its columns are isomorphisms, because ${}_RU^{(A)} \cong {}_RT(S^{(A)})$ and ${}_RU^{(B)} \cong {}_RT(S^{(B)})$. Thus it follows that ${}_RX \cong {}_RT(\mathrm{Coker}(H(g)))$, which implies that $\rho(X)$ is an isomorphism.

We are now in a position to prove the following refinement of Fuller [5, Theorem 2.6]:

<u>Theorem 10</u>. Let ${}_RU_S$ be a bimodule, and let $\rho(X): {}_RU \otimes_S \mathrm{Hom}_R(U, X) \longrightarrow {}_RX$ and $\sigma(Y): {}_SY \longrightarrow {}_S\mathrm{Hom}_R(U, U \otimes_S Y)$ be the canonical homo-

morphism for each left R-module X and left S-module Y. Let $\text{Gen}_R(U)$ be the class of those left R-modules which are sums of homomorphic images of ${}_RU$. Then the following conditions are equivalent:

(1) $\sigma(Y)$ is an isomorphism for all left S-modules Y and $\text{Gen}_R(U)$ is closed under submodules.

(2) $\sigma(Y)$ is an isomorphism for all left S-modules Y and $\rho(X)$ is a monomorphism for all left R-modules X.

(3) U_S is a weak generator (i.e. $U \otimes_S Y = 0$ for a left S-module Y implies Y = 0) and $\rho(X)$ is a monomorphism for all left R-modules X.

(4) ${}_RU$ is finitely generated quasi-projective, $S = \text{End}_R(U)$ and $\text{Gen}_R(U)$ is closed under submodules.

Proof. That (1) implies (2) is an immediate consequence of Proposition 9, while that (2) implies (3) follows from the obvious fact that if $\sigma(Y)$ is a monomorphism for all left S-modules Y then U_S is a weak generator.

Assume now (3). Then ${}_S\text{Hom}_R(U, U^{(A)}) \cong {}_SS^{(A)}$ canonically for all sets A and ${}_RU$ is Σ-quasi-projective by Corollaries 6 and 7, and thus ${}_RU$ is finitely generated quasi-projective and $S = \text{End}_R(U)$ by Proposition 8. Applying then Proposition 4, we know that U_S is flat, and, applying further Proposition 2, we conclude that $\text{Gen}_R(U)$ is closed under submodules. Thus we have the condition (4).

Assume next (4). Then we know again by Proposition 8 that ${}_RU$ is Σ-quasi-projective and ${}_S\text{Hom}_R(U, U^{(A)}) \cong {}_SS^{(A)}$ canonically for all sets A. Therefore it follows from Sato [5, Theorem 2.1] that $\sigma(Y)$ is an isomorphism for all left S-modules Y, because the Σ-quasi-projectivity of ${}_RU$ implies the semi-Σ-quasi-projectivity of ${}_RU$ in the sense that the functor $\mathrm{Hom}_R(U, -)$ preserves the exactness of sequences of the form ${}_RU^{(A)} \longrightarrow {}_RU^{(B)} \longrightarrow {}_RX \longrightarrow 0$. Thus we have the condition (1), and this completes the proof of our theorem.

Remark. It is proved in Fuller [1, Lemma 2.2] or Sato [5, Lemma 3.2] that if ${}_RU$ is quasi-projective and every submodule of ${}_RU$ is in $\mathrm{Gen}_R(U)$ then $\mathrm{Gen}_R(U)$ is closed under submodules. This shows that we can replace in (4) of Theorem 10 the last condition that "$\mathrm{Gen}_R(U)$ is closed under submodules" by the condition that "every submodule of ${}_RU$ is a sum of homomorphic images of ${}_RU$", obtaining thus the condition (e) of [1, Theorem 2.6] or the condition (5) of [5, Theorem 3.1]. We want now to point out that the above mentioned lemma of Fuller (and Sato) remains true even if we assume that ${}_RU$ is semi-Σ-quasi-projective instead of quasi-projective. For, given an epimorphism $h: {}_RU \longrightarrow {}_RX$, Ker(h) is in $\mathrm{Gen}_R(U)$ and therefore we have an exact sequence ${}_RU^{(A)} \longrightarrow {}_RU \longrightarrow {}_RX \longrightarrow 0$ for a suitable set A. The semi-Σ-quasi-projectivity of ${}_RU$ then implies that $\mathrm{Hom}_R(U, h)$ is an epimorphism, so that ${}_RU$ is quasi-projective. Since however if $\sigma(Y)$ is an isomorphism for all left S-modules Y then ${}_RU$ is semi-Σ-quasi-projective by [5, Theorem 2.1], we can conclude that the same replacement of conditions for ${}_RU$ as for (4) of Theorem 10 above is also available for (1) of Theorem 10.

In this connection, it is further to be noted that <u>if we assume that ${}_RU$ is semi-Σ-quasi-projective and every submodule of ${}_RU$</u>

<u>is in $Gen_R(U)$ then $_RU$ is not only quasi-projective but also Σ-quasi-projective</u>, as a matter of fact. For, given any epimorphism $h: {}_RU^{(B)} \longrightarrow {}_RX$, Ker(h) is in $Gen_R(U)$, since $Gen_R(U)$ is closed under submodules, so that we have an exact sequence ${}_RU^{(A)} \longrightarrow {}_RU^{(B)} \xrightarrow{h} {}_RX \longrightarrow 0$. Since $_RU$ is semi-Σ-quasi-projective, it follows then that $Hom_R(U, h)$ is an epimorphism.

This Research was supported by NSF under Grant MCS77-01756.

REFERENCES

1. K. R. Fuller, Density and Equivalence, J. of Algebra 29 (1974), 528-550.

2. K. R. Fuller and D. A. Hill, On quasi-projective modules via relative projectivity, Arch. Math. 21 (1970), 369-373.

3. J. Lambek, "Lectures on Rings and Modules", Blaisdell, Waltham, Mass., 1966.

4. E. de Robert, Projectifs et injectifs relatifs, C. R. Acad. Sci. Paris Ser. A 286 (1969), 361-364.

5. M. Sato, Fuller's theorem on equivalence, to appear in J. of Algebra.

ON INVERSIVE LOCALIZATION

John A. Beachy
Northern Illinois University
DeKalb, Illinois

P.M.Cohn introduced in [4] the inversive localization at a semiprime ideal N of a left Noetherian ring R. He gave a construction for a ring of quotients $R_{\Gamma(N)}$ universal with respect to the property that every matrix regular modulo N is invertible over $R_{\Gamma(N)}$. That is, in each ring $(R_{\Gamma(N)})_n$ of $n \times n$ matrices over $R_{\Gamma(N)}$, every element of $(R)_n$ which is regular modulo $(N)_n$ becomes invertible. The ring $R_{\Gamma(N)}$ always exists, but it can be very difficult to determine. In fact, it is hard to compute even the kernel of the mapping $R \to R_{\Gamma(N)}$. On the other hand, $R_{\Gamma(N)}$ has some very desirable properties which are lacking in the torsion theoretic localization $R_{C(N)}$, and so it appears to be worthy of further study. This paper contains the announcement of some preliminary results in studying inversive localization. It also contains some explicit computations, since one of the first tasks must be to build a collection of examples. Included is the computation of $\Lambda_{\Gamma(\Pi)}$ for every prime ideal Π of the ring of formal matrices $\Lambda = \begin{pmatrix} R & M \\ N & R \end{pmatrix}$, where R is a commutative ring and M and N are modules over R which have the pairings necessary to define matrix multiplication in Λ. This includes as special cases several examples given by Cohn in [4].

§1. Some properties of the inversive localization

The ring R is assumed to be an associative ring with identity, and all modules are assumed to be unital. If R is left Noetherian and N is a semiprime ideal of R, then the ring $R_{\Gamma(N)}$ is constructed as follows (see [4] and [5, p.255] for details). Let $\Gamma(N)$ be the set of all square matrices over R which are regular modulo N. For each $n \times n$ matrix $\gamma = (a_{ij}) \in \Gamma(N)$ take a set of n^2 symbols $(a'_{ij}) = \gamma'$, and take a ring presentation of $R_{\Gamma(N)}$ consisting of all of the elements of R, as well as all of the elements a'_{ij} as generators; as defining relations take all of the relations holding in R, together with the relations, in matrix form, $\gamma\gamma' = \gamma'\gamma = 1$, for each $\gamma \in \Gamma(N)$. The mapping $\lambda: R \to R_{\Gamma(N)}$ is an epimorphism in the category of rings, and $R_{\Gamma(N)}/J(R_{\Gamma(N)})$ is the classical ring of quotients of R/N, under the embedding $\lambda': R/N \to R_{\Gamma(N)}/J(R_{\Gamma(N)})$ induced by λ. (The Jacobson radical of the ring R will be denoted by $J(R)$.) The latter property will be used in Theorem 1.1 to characterize $R_{\Gamma(N)}$.

The ring $R_{\Gamma(N)}$ can be constructed in certain cases even when R is not left Noetherian. In fact, Cohn's proofs remain valid when N is any semiprime ideal such that the factor ring R/N is a left Goldie ring (this ensures the existence of the classical ring of quotients $Q_{c\ell}(R/N)$). A semiprime (prime) ideal which satisfies this condition will be called a semiprime (prime) Goldie ideal. Working in this generality means that the inversive localization can be defined, for example, at any prime ideal of a ring with polynomial identity.

If N is a semiprime Goldie ideal of the ring R, consider the following conditions on a ring S and ring homomorphism $\phi: R \to S$. Note that any ring which satisfies these conditions must be unique (up to isomorphism).

J_1. The homomorphism ϕ induces a ring homomorphism $\phi': R/N \to S/J(S)$ such that the following diagram commutes. (The mappings $R \to R/N$ and $S \to S/J(S)$ are the natural projections.)

$$\begin{array}{ccc} R & \xrightarrow{\phi} & S \\ \downarrow & & \downarrow \\ R/N & \xrightarrow{\phi'} & S/J(S) \end{array}$$

J_2. The ring $S/J(S)$ is a classical ring of quotients of R/N, under the embedding $\phi': R/N \to S/J(S)$.

J_3. If $\theta: R \to T$ is a ring homomorphism which satisfies conditions J_1 and J_2, then there exists a unique ring homomorphism $\theta^*: S \to T$ such that the following diagram commutes.

$$\begin{array}{ccc} R & \xrightarrow{\phi} & S \\ & {\scriptstyle\theta}\searrow & \downarrow {\scriptstyle\theta^*} \\ & & T \end{array}$$

THEOREM (1.1). Let N *be a semiprime Goldie ideal of* R. *Then the inversive localization* $\lambda: R \to R_{\Gamma(N)}$ *of* R *at* N *can be defined, and it satisfies conditions* J_1, J_2, *and* J_3.

The next theorem was proved by Cohn for Noetherian rings. Let C(N) denote the set of elements of R which are regular modulo N. The torsion theoretic localization at N, which is determined by C(N) when N is a semiprime Goldie ideal, will be denoted by $R_{C(N)}$. (See [3] and [6] for details.) Recall that C(N) is said to be a left denominator set if (i) for each $a \in R$ and $c \in C(N)$ there exist $a_1 \in R$ and $c_1 \in C(N)$ such that $c_1 a = a_1 c$ and (ii) if $ac = 0$ for $a \in R$ and $c \in C(N)$, then there exists $c_1 \in C(N)$ such that $c_1 a = 0$. If C(N) is a left denominator set, then $R_{C(N)}$ is a classical ring of left fractions of R, obtained by inverting the elements of C(N), and in this case it will be denoted by R_N.

THEOREM (1.2). Let N *be a semiprime Goldie ideal of* R. *Then the ring* $R_{\Gamma(N)}$ *is naturally isomorphic to the ring* $R_{C(N)}$ *if and only if* C(N) *is a left denominator set.*

Since the construction of $R_{\Gamma(N)}$ is left-right symmetric, it can sometimes be constructed in this manner even when it differs from $R_{C(N)}$. For example, let R be the ring $\begin{pmatrix} Z & 0 \\ Z & Z \end{pmatrix}$ of lower triangular matrices over the ring of integers Z, with the prime ideal $P = \begin{pmatrix} Z & 0 \\ Z & pZ \end{pmatrix}$, where $p \in Z$ is prime. It is not difficult to check that C(P) is a right denominator set but not a left denominator set. (This also follows from Proposition 2.2.) Since $\begin{pmatrix} m & 0 \\ n & 0 \end{pmatrix}\begin{pmatrix} 0 & 0 \\ 0 & 1 \end{pmatrix} = 0$ for any $m, n \in Z$, the ideal $\begin{pmatrix} Z & 0 \\ Z & 0 \end{pmatrix}$ must be in the kernel of $\lambda: R \to R_{\Gamma(P)}$, and computing the classical ring of right fractions shows that $R_{\Gamma(P)}$ is just the localization $Z_{(p)}$ of Z at pZ. A computation of the torsion theoretic localization $R_{C(P)}$ (on the left) shows it to be the full ring of 2×2 matrices over $Z_{(p)}$.

For left Artinian rings it has been possible to explicitly compute $R_{\Gamma(N)}$. It turns out to be just a homomorphic image of R, and is in fact the largest homomorphic image in which C(N) becomes a left denominator set.

THEOREM (1.3). Let N *be a semiprime ideal of the left Artinian ring* R. *Then* $R_{\Gamma(N)} = R/N^k$, *where* $N^k = N^{k+1} = \dots$.

Theorem 1.3 follows from part (c) of the next proposition, which has been helpful in computing $R_{\Gamma(N)}$ in a number of examples. The first two parts follow immediately from Theorem 3.2 of [4]. The proof of part (c) has been included since it illustrates some of the techniques which must be used.

PROPOSITION (1.4). Let N *be a semiprime Goldie ideal of* R, *let* K *be the kernel of the homomorphism* $\lambda : R \to R_{\Gamma(N)}$, *and let* I *be an ideal of* R *which is contained in* N.

(a). If $I \subseteq K$, *then* $(R/I)_{\Gamma(N/I)} = R_{\Gamma(N)}$.

(b). If $I \subseteq K$ *and* C(N) *is a left denominator set modulo* I, *then* $R_{\Gamma(N)} = (R/I)_{N/I}$.

(c). If $I = I^2$ *and* I *is finitely generated either as a left or as a right ideal, then* $I \subseteq K$.

Proof. (c) Assume that $I = \Sigma_{i=1}^{n} Rx_i$, for $x_1, \dots, x_n \in I$. If $I = I^2$, then $I = \Sigma_{i=1}^{n} Ix_i$, and so $x_i = \Sigma_{j=1}^{n} a_{ij}x_j$, for $a_{ij} \in I$. In matrix form, this shows that $(1-\gamma)u = 0$ for the matrix $\gamma = (a_{ij})$ and the vector u which has entries $x_1, \dots, x_n$. Since $1-\gamma \equiv 1 \pmod N$, $1-\gamma \in \Gamma(N)$, and so $1-\gamma$ must be invertible over $R_{\Gamma(N)}$. Therefore $x_i \in K$, for each i, and $I \subseteq K$.

§2. Examples

Let R and S be associative rings with identity, and let ${}_R M_S$ and ${}_S N_R$ be unital bimodules. Let Λ be the ring of 2×2 matrices $\begin{pmatrix} R & M \\ N & S \end{pmatrix}$. To define a multiplication for the ring Λ it is necessary to have a Morita context (see [1]). That is, it is necessary to have bilinear mappings $(\, , \,) : M \otimes_S N \to R$ and $[\, , \,] : N \otimes_R M \to S$, together with associative laws $m_1[n,m] = (m_1,n)m$ and $[n,m]n_1 = n(m,n_1)$, which must hold for all $m, m_1 \in M$ and $n, n_1 \in N$. Some elementary facts about Λ must be given, at the risk of writing down results which are in the folklore of the subject.

If I is an ideal of Λ, then I must have the form $I = \begin{pmatrix} A & X \\ Y & B \end{pmatrix}$, where A and B are ideals of R and S, respectively, and ${}_RX_S$, ${}_SY_R$ are submodules of M and N, respectively. Furthermore, the following conditions must hold.

$$\begin{array}{llll} (M,Y) \subseteq A & AM \subseteq X & NA \subseteq Y & [Y,M] \subseteq B \\ (X,N) \subseteq A & MB \subseteq X & BN \subseteq Y & [N,X] \subseteq B \end{array}$$

From this point on, it seems to be much the easiest to suppress all mention of the bilinear mappings (,) and [,], except in the statements of theorems.

The above characterization of ideals can be used to show that if I is any ideal with $I \cap R = A$, then

$$\begin{pmatrix} A & AM \\ NA & NAM \end{pmatrix} \subseteq I \subseteq \begin{pmatrix} A & AN^{-1} \\ M^{-1}A & M^{-1}AN^{-1} \end{pmatrix}$$

where $AN^{-1} = \{x \in M \mid xN \subseteq A\}$, $M^{-1}A = \{y \in N \mid My \subseteq A\}$ and $M^{-1}AN^{-1} = \{b \in S \mid MbN \subseteq A\}$. Similarly, if $I \cap M = X$, then

$$\begin{pmatrix} XN & X \\ NXN & NX \end{pmatrix} \subseteq I \subseteq \begin{pmatrix} XM^{-1} & X \\ M^{-1}XM^{-1} & M^{-1}X \end{pmatrix}$$

where $XM^{-1} = \{a \in R \mid aM \subseteq X\}$, $M^{-1}X = \{b \in S \mid Mb \subseteq X\}$ and $M^{-1}XM^{-1} = \{y \in N \mid MyM \subseteq X\}$. Similar conditions can be given in the other two cases.

PROPOSITION (2.1). If $\Pi = \begin{pmatrix} P & X \\ Y & Q \end{pmatrix}$ *is a prime ideal of* Λ, *then* P *and* Q *are prime ideals (if proper). Furthermore,* Π *must be one of the following types.*

Type 1. If $(M,N) \subseteq P$, *then* $[N,M] \subseteq Q$, $X = M$, $Y = N$ *and either* $Q = S$ *or* $P = R$.

Type 2. If $(M,N) \not\subseteq P$, *then* $[N,M] \not\subseteq Q$, $X \neq M$ *and* $Y \neq N$.

Proof. If A and B are ideals of R with $AB \subseteq P$, then for the left ideals generated by A and B, $\begin{pmatrix} A & 0 \\ NA & 0 \end{pmatrix}\begin{pmatrix} B & 0 \\ NB & 0 \end{pmatrix} = \begin{pmatrix} AB & 0 \\ NAB & 0 \end{pmatrix} \subseteq \Pi$, so $A \subseteq P$ or $B \subseteq P$. Similarly, if Q is proper, then it is a prime ideal.

If $MN \subseteq P$, then $I^2 \subseteq \Pi$ for the left ideal $I = \begin{pmatrix} MN & M \\ N & NM \end{pmatrix}$, and so $I \subseteq \Pi$. Then $\begin{pmatrix} R & 0 \\ N & 0 \end{pmatrix}\begin{pmatrix} 0 & M \\ 0 & S \end{pmatrix} \subseteq \Pi$, which shows that either $Q = S$ or $P = R$.

On the other hand, if $MN \not\subseteq P$, then by the above argument $NM \not\subseteq Q$. The conditions which Π satisfies (as an ideal) force $X \neq M$ and $Y \neq N$.

As the proof of Proposition 2.1 shows, semiprime ideals can be treated in a similar manner. The next proposition determines the inversive localization at a prime ideal of Type 1. It also shows that the set of elements regular modulo such a prime ideal need not be a left denominator set, even when R is commutative and $R = S$.

PROPOSITION (2.2). Let $(M,N) = I$, *let* P *be a prime Goldie ideal of* R *with* $P \supseteq I$, *and let* Π *be the prime Goldie ideal* $\begin{pmatrix} P & M \\ N & S \end{pmatrix}$ *of* Λ.

(a). $\Lambda_{\Gamma(\Pi)} = (R/I)_{\Gamma(P/I)}$.

(b). $C(\Pi)$ *is a left denominator set if and only if* $C(P)$ *is a left denominator set and* ${}_RM$ *is* $C(P)$*-torsion.*

Proof. (a) Since $\begin{pmatrix} 1 & 0 \\ 0 & 0 \end{pmatrix} \in C(\Pi)$, it follows that the kernel of $\lambda : \Lambda \to \Lambda_{\Gamma(\Pi)}$ must contain $\begin{pmatrix} MN & M \\ N & S \end{pmatrix}$, since $\begin{pmatrix} 1 & 0 \\ 0 & 0 \end{pmatrix}\begin{pmatrix} 0 & 0 \\ N & S \end{pmatrix} = 0$ and $\begin{pmatrix} 0 & M \\ 0 & S \end{pmatrix}\begin{pmatrix} 1 & 0 \\ 0 & 0 \end{pmatrix} = 0$. The inversive localization can then be computed by using Proposition 1.4 (a).

(b) =>) If $m \in M$, then $\begin{pmatrix} 0 & m \\ 0 & 0 \end{pmatrix}\begin{pmatrix} 1 & 0 \\ 0 & 0 \end{pmatrix} = 0$, so there must exist an element $\begin{pmatrix} c & x \\ y & b \end{pmatrix} \in C(\Pi)$ with $\begin{pmatrix} c & x \\ y & b \end{pmatrix}\begin{pmatrix} 0 & m \\ 0 & 0 \end{pmatrix} = 0$. Thus $cm = 0$ for some $c \in C(P)$, and M is $C(P)$-torsion. It is just as easy to show that $C(P)$ must be a left denominator set.

<=) Let $C^{\Delta}(\Pi)$ denote the elements of $C(\Pi)$ of the form $\begin{pmatrix} c & 0 \\ 0 & 0 \end{pmatrix}$ for $c \in C(\Pi)$. If $\gamma = \begin{pmatrix} c & x \\ y & d \end{pmatrix} \in C(\Pi)$, then $c_1 x = 0$ for some $c_1 \in C(P)$, since M is $C(P)$-torsion, and then $\gamma^*\gamma \in C^{\Delta}(\Pi)$, for $\gamma^* = \begin{pmatrix} c_1 & 0 \\ 0 & 0 \end{pmatrix} \in C^{\Delta}(\Pi)$. Given $\alpha = \begin{pmatrix} a & w \\ z & b \end{pmatrix} \in \Lambda$, by assumption it is possible to find $a_1 \in R$ and $c_2 \in C(P)$ with $c_2 a = a_1 c_1 c$. In addition there exists $c_3 \in C(P)$ with $c_3 w = 0$, so setting $\gamma_1 = \begin{pmatrix} c_3 c_2 & 0 \\ 0 & 0 \end{pmatrix}$, $\alpha_1 = \begin{pmatrix} c_3 a_1 & 0 \\ 0 & 0 \end{pmatrix}$ gives $\gamma_1\alpha = (\alpha_1\gamma^*)\gamma$. If $\alpha\gamma = 0$, then there exist $\alpha_1 \in \Lambda$ and $\gamma_1 \in C^{\Delta}(\Pi)$ with $\gamma_1\alpha = \alpha_1\gamma^*$. Thus $\alpha_1(\gamma^*\gamma) = \gamma_1(\alpha\gamma) = 0$. Since $C(P)$ is a left denominator set, it is easy to find $\gamma_2 \in C^{\Delta}(\Pi)$ with $\gamma_2\alpha_1 = 0$, and then $(\gamma_2\gamma_1)\alpha = \gamma_2(\alpha_1\gamma^*) = 0$.

A module ${}_RM$ is said to be prime if $AX \neq 0$ for all nonzero ideals $A \subseteq R$ and all nonzero submodules $X \subseteq M$. Similarly, M is said to be semiprime if $AX = 0$ implies $AM \cap X = 0$, for all ideals A and submodules X. Semiprime ideals of Type 2 can be characterized in a manner similar to that of the following proposition, with semiprime ideals and modules instead of prime ideals and modules. The third condition must be replaced by the condition that for submodules $W \subseteq M$ and $Z \subseteq N$,

$WNW \subseteq X$ implies $W \subseteq X$ and $ZMZ \subseteq Y$ implies $Z \subseteq Y$. Note that the conditions of the proposition are symmetric, in the sense that the conditions could have been given in terms of $(M/X)_S$, $_S(N/Y)$ and $(\ ,\ )$.

PROPOSITION (2.3). Let $\Pi = \begin{pmatrix} P & X \\ Y & Q \end{pmatrix}$ *be an ideal of* Λ, *with prime ideals* $P \not\supseteq (M,N)$ *and* $Q \not\supseteq [N,M]$ *and submodules* $X \neq M$ *and* $Y \neq N$. *The following conditions are equivalent.*

(1). Π *is a prime ideal.*

(2). $_R(M/X)$ *and* $(N/Y)_R$ *are prime modules over* R/P.

(3). For $m \in M$ *and* $n \in N$, $[N,m] \subseteq Q$ *implies* $m \in X$ *and* $[n,M] \subseteq Q$ *implies* $n \in Y$.

Proof. (1) => (2) If $_RA \subseteq R$ and $_RW \subseteq M$ with $AW \subseteq X$, then $\begin{pmatrix} A & 0 \\ NA & 0 \end{pmatrix}\begin{pmatrix} 0 & W \\ 0 & NW \end{pmatrix} = \begin{pmatrix} 0 & AW \\ 0 & NAW \end{pmatrix} \subseteq \begin{pmatrix} P & X \\ Y & Q \end{pmatrix}$. Thus either $A \subseteq P$ or $W \subseteq X$, and so M/X is a prime R/P-module. Similarly, $(N/Y)_R$ is a prime R/P-module.

(2) => (3) If $m \in M$ and $Nm \subseteq Q$, then $MN(Rm) \subseteq MQ \subseteq X$, and so $Rm \subseteq X$. Similarly, $nM \subseteq Q$ implies $n \in Y$.

(3) => (1) Suppose that $\begin{pmatrix} A & W \\ Z & B \end{pmatrix}\begin{pmatrix} C & U \\ V & D \end{pmatrix} \subseteq \begin{pmatrix} P & X \\ Y & Q \end{pmatrix}$ for two ideals of Λ. Then $BD \subseteq Q$ implies that $B \subseteq Q$ or $D \subseteq Q$, since Q is prime, so suppose that $B \subseteq Q$. Then $NW \subseteq B \subseteq Q$ implies that $W \subseteq X$ and $ZM \subseteq B \subseteq Q$ implies that $Z \subseteq Y$. Finally, $NA \subseteq Z \subseteq Y$ implies that $(MN)A \subseteq MY \subseteq P$, which implies that $A \subseteq P$ since $MN \not\subseteq P$.

At the beginning of this section it was remarked that if I is any ideal of Λ such that $I \cap R = P$, then $I \subseteq \begin{pmatrix} P & PN^{-1} \\ M^{-1}P & M^{-1}PN^{-1} \end{pmatrix}$. As can be seen from the conditions of the proposition, $X = PN^{-1}$, $Y = M^{-1}P$, and $Q = M^{-1}PN^{-1}$, so $I \subseteq \Pi$. Thus any prime ideal of Type 2 contains all ideals which "lie over" any of its components. It also follows from the proposition that if $cm \in X$ for $c \in C(P)$ and $m \in M$, then $m \in X$. (If $cm \in X$, then $c(mN) \subseteq XN \subseteq P$ implies $mN \subseteq P$, so $m \in X$.)

The final proposition shows that if R and S are commutative, then the inversive localization is defined at any prime ideal of Λ. In this situation I have been unable to determine the inversive localization at prime ideals of Type 2, unless the denominator set conditions are satisfied, so it remains an open question. If $S = R$, (still commutative) then Q must be equal to P since it is the annihilator of M/X, and it follows immediately that the denominator set conditions must be satisfied. I have been told by Bill Blair that part (a) of Proposition 2.4 has been known to Lance Small for several years, for semiprime ideals. (With some slight modifications the proof given below can be used to obtain the more general result.)

PROPOSITION (2.4). Let $\Pi = \begin{pmatrix} P & X \\ Y & Q \end{pmatrix}$ *be a prime ideal of* Λ, *with* $(M,N) \not\subseteq P$. *Assume that* P *and* Q *are prime ideals such that* R/P *and* S/Q *are left Goldie rings.*

(a). Λ/Π *is a left Goldie ring.*

(b). $Q_{cl}(\Lambda/\Pi) = \begin{pmatrix} Q_{cl}(R/P) & Q_{cl}(R/P)\otimes_R(M/X) \\ Q_{cl}(S/Q)\otimes_S(N/Y) & Q_{cl}(S/Q) \end{pmatrix}$

(c). $C(\Pi)$ *is a left denominator set if and only if* $C(P)$ *and* $C(Q)$ *are left denominator sets and the following conditions hold.*

(i) For $c \in C(P)$, $d \in C(Q)$, $m \in M$, $n \in N$, *there exist* $c_1 \in C(P)$, $d_1 \in C(Q)$, $m_1 \in M$, $n_1 \in N$ *with* $c_1m = m_1d$ *and* $d_1n = n_1c$.

(ii) If $m \in M$ *and* $n \in N$ *with* $md = 0$ *and* $nc = 0$ *for* $d \in C(Q)$, $c \in C(P)$, *then there exist* $c_1 \in C(P)$ *and* $d_1 \in C(Q)$ *with* $c_1m = 0$ *and* $d_1n = 0$.

(d). If $C(\Pi)$ *is a left denominator set, then*

$$\Lambda_\Pi = \begin{pmatrix} R_P & R_P\otimes_R M \\ S_Q\otimes_S N & S_Q \end{pmatrix}$$

Proof. (a) By Theorem 2.2 of [2], a prime ring R is a left Goldie ring if and only if it contains a uniform left ideal and each nonzero ideal contains a set of elements $a_1,\dots,a_n$ for which the left annihilator $\ell(a_1,\dots,a_n)$ is zero. Assume (without loss of generality) that $\Pi = 0$. If U is a uniform left ideal of R, then $\begin{pmatrix} U & 0 \\ NU & 0 \end{pmatrix}$ is a uniform left ideal of Λ. This follows from the observation that for any left ideal $\begin{pmatrix} A & 0 \\ Z & 0 \end{pmatrix} \subseteq \begin{pmatrix} U & 0 \\ NU & 0 \end{pmatrix}$, $NZ \subseteq A$ is nonzero if Z is nonzero. For any ideal $\begin{pmatrix} A & X \\ Y & B \end{pmatrix}$ which is nonzero, both A and B must be nonzero, so there exist elements $a_1,\dots,a_n \in A$ and $b_1,\dots,b_k \in B$ with $\ell_R(a_1,\dots,a_n) = 0$ and $\ell_S(b_1,\dots,b_k) = 0$. Using these elements it is easy to construct a finite set of matrices in $\begin{pmatrix} A & X \\ Y & B \end{pmatrix}$ with zero left annihilator. (If, for the left annihilator $\begin{pmatrix} C & W \\ Z & D \end{pmatrix}$, it can be shown that $C = 0$ and $D = 0$, then $MZ = 0$ and $NW = 0$ imply that $Z = 0$ and $W = 0$.)

(b) This will follow from (a) and (d).

(c) Before beginning the proof of (c) it is necessary to establish some properties of $C(\Pi)$. If $\gamma \in C(\Pi)$, then there exists $\gamma' \in C(\Pi)$ such that $\gamma'\gamma \in C(\Pi)$ and $\gamma'\gamma$ is a diagonal matrix modulo Π. Thus $\gamma'\gamma$ must have elements of $C(P)$ and $C(Q)$, respectively, along the diagonal. It suffices to show this when $\Pi = 0$. Since any regular element of a prime Goldie ring generates an essential left ideal, it follows that $\begin{pmatrix} A & 0 \\ Y & 0 \end{pmatrix} = \Lambda\gamma \cap \begin{pmatrix} R & 0 \\ N & 0 \end{pmatrix}$ is essential in $\begin{pmatrix} R & 0 \\ N & 0 \end{pmatrix}$. It can be checked that A must be an essential left ideal of R, so A contains a regular element. A similar argument can be given in S, and so $\Lambda\gamma$ contains an element $\begin{pmatrix} c & 0 \\ 0 & d \end{pmatrix} = \gamma'\gamma$, where $c \in C(P)$, $d \in C(Q)$. If $\alpha\gamma' = 0$, then $\alpha(\gamma'\gamma) = 0$, so $\alpha = 0$ and γ' is left regular, which implies that γ' is regular.

To prove (c), first assume that $C(\Pi)$ is a left denominator set. It is not difficult to show that $C(P)$ and $C(Q)$ are left denominator sets, by using the preceding remark and the fact that $\begin{pmatrix} c & 0 \\ 0 & d \end{pmatrix} \in C(\Pi)$ if $c \in C(P)$ and $d \in C(Q)$. Given $m \in M$ and $d \in C(Q)$, there must exist $\begin{pmatrix} a_1 & m_1 \\ n_1 & b_1 \end{pmatrix} \in \Lambda$ and $\begin{pmatrix} c_1 & x_1 \\ y_1 & d_1 \end{pmatrix} \in C(\Pi)$ with $\begin{pmatrix} a_1 & m_1 \\ n_1 & b_1 \end{pmatrix}\begin{pmatrix} 1 & 0 \\ 0 & d \end{pmatrix} = \begin{pmatrix} c_1 & x_1 \\ y_1 & d_1 \end{pmatrix}\begin{pmatrix} 0 & m \\ 0 & 0 \end{pmatrix}$. This implies that $m_1 d = c_1 m$, and it can be assumed, as above, that $c_1 \in C(P)$. The other condition can be proved similarly.

Conversely, assume that the stated conditions hold. Let $C^\Delta(\Pi)$ be the set of elements of $C(\Pi)$ of the form $\begin{pmatrix} c & 0 \\ 0 & d \end{pmatrix}$. If $\gamma \in C(\Pi)$, then there exists $\gamma' \in C(\Pi)$ with $\gamma'\gamma = \begin{pmatrix} c & m \\ n & d \end{pmatrix}$, where $c \in C(P)$ and $d \in C(Q)$. By assumption there exist $c_1 \in C(P)$ and $m_1 \in M$ with $c_1 m = m_1 d$. Thus $\begin{pmatrix} c_1 & -m_1 \\ 0 & 1 \end{pmatrix}\begin{pmatrix} c & m \\ n & d \end{pmatrix} = \begin{pmatrix} c_1c-m_1n & 0 \\ n & d \end{pmatrix}$, and it can be checked that $\begin{pmatrix} c_1 & -m_1 \\ 0 & 1 \end{pmatrix} \in C(\Pi)$. The argument can be repeated for the entry n, so that the following result holds. If $\gamma \in C(\Pi)$, then there exists $\gamma^* \in C(\Pi)$ such that $\gamma^*\gamma \in C^\Delta(\Pi)$.

Now suppose that $\alpha \in \Lambda$ and $\gamma \in C(\Pi)$, with $\gamma^*\gamma = \begin{pmatrix} c & 0 \\ 0 & d \end{pmatrix} \in C^\Delta(\Pi)$ and $\alpha = \begin{pmatrix} a & x \\ y & b \end{pmatrix}$. Using the given conditions, it is possible to find $c_1 \in C(P)$, $d_1 \in C(Q)$ and $\alpha_1 = \begin{pmatrix} a_1 & x_1 \\ y_1 & b_1 \end{pmatrix} \in \Lambda$ such that $c_1 a = a_1 c$, $c_1 x = x_1 d$, $d_1 y = y_1 c$ and $d_1 b = b_1 d$. Thus $\gamma_1 \alpha = (\alpha_1 \gamma^*)\gamma$, for $\gamma_1 = \begin{pmatrix} c_1 & 0 \\ 0 & d_1 \end{pmatrix} \in C^\Delta(\Pi)$. If $\alpha\gamma = 0$, then by the above computations there exist $\alpha_1 = \begin{pmatrix} a_1 & x_1 \\ y_1 & b_1 \end{pmatrix} \in \Lambda$ and $\gamma_1 \in C^{\Delta}(\Pi)$ with $\gamma_1\alpha = \alpha_1\gamma^*$. Thus $\alpha_1(\gamma^*\gamma) = \gamma_1(\alpha\gamma) = 0$, so it follows that $a_1c = 0$, $x_1d = 0$, $y_1c = 0$ and $b_1d = 0$. Using the given conditions it is possible to find $\gamma_2 \in C^{\Delta}(\Pi)$ with $\gamma_2\alpha_1 = 0$. Thus $(\gamma_2\gamma_1)\alpha = (\gamma_2\alpha_1)\gamma^* = 0$.

(d) Assume that $C(\Pi)$ is a left denominator set. If $\alpha \in \Lambda$ and $\gamma\alpha = 0$ for $\gamma \in C(\Pi)$, then $\gamma^*\gamma\alpha = 0$, where $\gamma^*\gamma \in C^{\Delta}(\Pi)$. This shows that the entries of α must belong to the appropriate torsion submodules, which will be denoted by $\tau(\ )$, so with this notation the $C(\Pi)$-torsion ideal of Λ is $\begin{pmatrix} \tau_P(R) & \tau_P(M) \\ \tau_Q(N) & \tau_Q(S) \end{pmatrix}$. It suffices to consider the case in which Λ has no $C(\Pi)$-torsion. The localization $M_P \cong R_P \otimes_R M$ can be given a right S_Q-module structure as follows. For $c^{-1}m \in M_P$ and $d^{-1}b \in S_Q$, let $(c^{-1}m)(d^{-1}b) = (c_1c)^{-1}m_1b$, where $c_1m = m_1d$ for $c_1 \in C(P)$ and $m_1 \in M$. In addition, a pairing can be defined for $c^{-1}m \in M_P$ and $d^{-1}n \in N_Q$ by setting $(c^{-1}m, d^{-1}n) = (c_1c)^{-1}(m_1,n) \in R_P$, where $c_1 \in C(P)$ and $m_1 \in M$ with $c_1m = m_1d$. This can be done formally by showing that $(R_P \otimes_R M) \otimes_S S_Q \cong R_P \otimes_R M$. In this way the set of matrices $\Omega = \begin{pmatrix} R_P & M_P \\ N_Q & S_Q \end{pmatrix}$ can be given a ring structure, with Λ identified in the natural way with a subring of Ω. If $\gamma \in C(\Pi)$, then there exists $\gamma^* \in C(\Pi)$ with $\gamma^*\gamma \in C^{\Delta}(\Pi)$. It is clear that $\gamma^*\gamma$ is invertible in Ω, so γ is left invertible, and hence invertible (γ is a regular element of Ω). If $\omega \in \Omega$, it is evident that $\gamma \in C^{\Delta}(\Pi)$ can be found with $\gamma\omega = \alpha \in \Lambda$, so $\omega = \gamma^{-1}\alpha$, and this shows that Ω is the left ring of fractions Λ_Π defined by $C(\Pi)$.

I would like to thank Bill Blair for several very helpful conversations.

REFERENCES

1. S.A.Amitsur, *Rings of quotients and Morita contexts*, J. Algebra 17(1971), 273-298.

2. J.A.Beachy and W.D.Blair, *Rings whose faithful left ideals are cofaithful*, Pacific J. Math 58(1975), 1-13.

3. ________, *Localization at semiprime ideals*, J. Algebra 38(1976), 309-314.

4. P.M.Cohn, *Inversive localization in Noetherian rings*, Comm. Pure Appl. Math. 26 (1973), 679-691.

5. ________, Free Rings and their Relations, Academic Press (1971).

6. B.Stenström, Rings of Quotients, Springer-Verlag (1970).

SEMICRITICAL MODULES AND k-PRIMITIVE RINGS

A. K. Boyle and E. H. Feller
University of Wisconsin-Milwaukee
Milwaukee, Wisconsin 53201

1. Introduction. Let $|M|$ denote the Krull dimension of a right module M. If R is a semiprime ring with Krull dimension, then $|R/L| < |R|$ for all large right ideals L of R. This condition, which is called the large condition, is satisfied by a larger class of modules and yields results which parallel the results for semiprime rings.

A right module M is termed semicritical provided there exists a finite collection of submodules $K_1,\ldots,K_n$ such that $\bigcap_{i=1}^{n} K_i = 0$ and M/K_i is a critical module for all i. If, in addition, M/K_i is compressible for all i, we call M semicompressible.

In section 2, we show that every semicritical module M has a critical composition series, where the length of M is equal to its uniform dimension, denoted by dim M. In addition, M is shown to satisfy the large condition. In fact, for smooth modules, the semicritical condition is equivalent to the large condition.

In section 3, we consider rings which have a finitely generated faithful critical module of Krull dimension k. Such rings are called k-primitive rings, while an ideal D of a ring R is called coprimitive provided R/D is k-primitive for some k. This is a departure from the definition in [3] when the faithful critical was required to be cyclic. We first show that this change in definition does not alter the result in [3] that the Krull radical is the intersection of the coprimitive ideals.

We then proceed to examine the natural relationship between semicritical rings and k-primitive rings R with $|R| = k$. As prime rings with Krull dimension are characterized as isotopic semicompressible rings, we characterize isotopic semicritical rings as k-primitive rings R where $|R| = k$. Further, it is shown that a ring R is of this type if and only if R has a finitely generated faithful nonsingular critical module. Properties of these rings are discussed in section 4.

Continuing the analogy with prime rings, the last section is devoted to the question of an artinian classical right quotient ring for k-primitive rings of Krull dimension k. When this quotient ring does exist, we have a structure which extends the results for prime rings.

Conventions. If L is a large submodule of a module M, we write $L \leq_e M$. If N is a submodule of M, then $cl(N) = \{x \in M \mid xL \subseteq N$, for some $L \leq_e R\}$. If $N = cl(N)$, then N is said to be closed. A submodule I of M is a complement submodule if it is maximal in the collection of submodules K such that $K \cap N = 0$ for some submodule N.

All rings will have identity, and all modules are unitary right modules. Ideal shall mean two sided ideal. We denote the injective hull of a module M by E(M), and the singular submodule by Z(M). If S is a subset of M_R then $S^r = \{x \in R \mid sx = 0$ for all $s \in S\}$. If K is a submodule of M such that M/K is critical, then K is called cocritical.

2. Semicritical Modules. A module M has a critical composition series if there exists a finite descending chain of submodules $M = M_o \supset M_1 \supset \ldots \supset M_n = 0$, where M_{i-1}/M_i is α_i-critical and $\alpha_1 \geq \alpha_2 \geq \ldots \geq \alpha_n$. In [8], [9,2.6] it is shown that the length of this chain is unique and we will denote this length by $\ell(M)$. In this section we consider properties of semicritical modules and their critical composition series.

2.1 Theorem. Let M_R be a semicritical module. Then

(1) M has Krull dimension.

(2) Every nonzero submodule has a critical composition series.

(3) If L is a large submodule of M, then $\ell(L) = \ell(M)$.

(4) Every nonzero submodule of M is semicritical.

Proof: (2). Since M is semicritical there exists a finite collection of cocritical submodules $K_1, \ldots, K_n$ such that $\bigcap_{i=1}^{n} K_i = 0$, where this intersection is irredundant. Number the K_i so that if $\alpha_i = |R/K_i|$, then $\alpha_1 \geq \alpha_2 \geq \ldots \geq \alpha_n$. Let N be a nonzero submodule of M and consider the chain

$$\text{(A)} \qquad N \supset N \cap K_1 \supset N \cap K_1 \cap K_2 \supset \ldots \supset N \cap K_1 \cap \ldots \cap K_n = 0.$$

If $N \cap K_1 \cap \ldots \cap K_j \neq N \cap K_1 \cap \ldots \cap K_{j+1}$, then $N \cap K_1 \cap \ldots \cap K_j/N \cap K_1 \cap \ldots \cap K_{j+1} \cong N \cap K_1 \cap \ldots \cap K_j + K_{j+1}/K_{j+1} \subseteq R/K_{j+1}$ and hence is critical of Krull dimension α_{j+1}. Thus after identifying equal members of the sequence (A) we have a critical composition series for N.

(1). This follows from (2) by using [7,1.1].

(3). Suppose L is a large submodule of M and suppose $L \cap K_1 \cap \ldots \cap K_j = L \cap K_1 \cap \ldots \cap K_{j+1}$. Then $L \cap K_1 \cap \ldots \cap K_j \cap K_{j+2} \cap \ldots \cap K_n \subseteq \bigcap_{i=1}^{n} K_i = 0$. However, L is large and $K_1 \cap \ldots \cap K_j \cap K_{j+2} \cap \ldots \cap K_n \neq 0$ since the intersection $\bigcap_{i=1}^{n} K_i$ is irredundant. This is a contradiction. Thus all the terms in the sequence (A) are distinct and hence $\ell(L) = \ell(M)$.

(4). Let N be a nonzero submodule of M. If $N \not\subseteq K_i$, then $N/K_i \cap N \cong K_i + N/K_i$ is critical. This implies N is semicritical.

2.2 Theorem. Let M be a semicritical module. If N is a nonzero submodule of M, then $\ell(N) = \dim N$.

Proof: By [7,2.2] there exists an essential direct sum $K = C_1 \oplus \ldots \oplus C_s$ of critical submodules of M. Let $\alpha_i = |C_i|$ and number the C_i so that $\alpha_1 \geq \alpha_2 \geq \ldots \geq \alpha_s$. Then $K = C_1 \oplus \ldots \oplus C_s \supset C_2 \oplus \ldots \oplus C_s \supset \ldots \supset C_s \supset 0$ is a critical composition series for K of length s. Since K is large, dim K = dim M and by 2.1, $\ell(M) = \ell(K) = s = \dim K = \dim M$. If N is any nonzero submodule, by 2.1, N is semicritical and therefore $\ell(N) = \dim N$.

2.3 Corollary. If M is a semicritical module, then every uniform submodule of M is critical.

Proof: If N is a uniform submodule of M, then $\ell(N) = \dim N = 1$. Thus N is necessarily critical.

A module M with Krull dimension is said to satisfy the large condition provided $|M/L| < |M|$ for any large submodule L of M. In [7,6.1] it is shown that every semiprime ring with Krull dimension satisfies the large condition. This theorem can be extended to semicritical modules.

2.4 Theorem. If M is semicritical, then M satisfies the large condition.

Proof: The proof is by induction on n = dim M. If n = 1, M is critical by 2.3 and the result follows directly. Assume the result is true for all semicritical modules of uniform dimension n - 1.

Suppose M is a semicritical module with dim M = n. Then there exists a finite irredundant intersection of cocritical submodules, say $0 = \bigcap_{i=1}^{n} K_i$, where by 2.2, $n = \ell(M) = \dim M$. Let $C_i = K_1 \cap \ldots \cap K_{i-1} \cap K_{i+1} \cap \ldots \cap K_n$. Since $C_i \cap K_i = 0$, C_i is critical for all i. Now suppose $L \leq_e M$ and $|M/L| = |M| = \alpha$. Then $|C_1 + L/L| = |C_1/C_1 \cap L| < \alpha$. Therefore, since $|M/L| = \sup\{|M/C_1 + L|, |C_1 + L/L|\}$, necessarily $|M/C_1 + L| = \alpha$. Continuing in this manner we obtain $|M/C_1 \oplus C_2 \oplus \ldots \oplus C_n + L| = \alpha$ and, consequently, $|M/C_1 \oplus \ldots \oplus C_n| = \alpha$.

Since $C_1 = \bigcap_{i=2}^{n} K_i$, M/C_1 is a semicritical module with uniform dimension n-1. By hypothesis, M/C_1 satisfies the large condition. Since $\dim(C_1 \oplus \ldots \oplus C_n/C_1) = n-1$, then $C_1 \oplus \ldots \oplus C_n/C_1 \leq_e M/C_1$ and $|M/C_1 \oplus \ldots \oplus C_n| = |M/C_1/C_1 \oplus \ldots \oplus C_n/C_1| < \alpha$, which is a contradiction.

From 2.4 the following results are true for semicritical modules.

2.5 Corollary. Let M be a module with Krull dimension satisfying the large condition. Let $\{E_i | i \in I\}$ be the collection of large submodules of M. Then $|M| = \sup\{|M/E_i| + 1 | i \in I\}$.

2.6 Corollary. Let R be a ring satisfying the large condition, where $|R| = \alpha$.

(1) Every α-critical is nonsingular.

(2) Every finitely generated uniform module U with $|U| = \alpha$ is critical.

(3) Every α-critical module contains an isomorphic copy of a right ideal of R.

Proof: (1). Suppose C is critical with $Z(C) \neq 0$. Then C contains a nonzero submodule $C_0 \cong R/I$ for some large right ideal I. Since R satisfies the large condition, $|C| = |C_0| = |R/I| < \alpha$.

(2). Let U be a finitely generated uniform right module, say $U = \sum_{i=1}^{n} x_i R$, with $|U| = \alpha$ and let U_0 be a nonzero submodule of U. Define $I_i = x_i^{-1} U_0 = \{r \in R \mid x_i r \in U_0\}$. Since $U_0 \leq_e U$, then $I_i \leq_e R$. Furthermore, $U/U_0 = \sum_{i=1}^{n} x_i R + U_0/U_0$, where $x_i R + U_0/U_0 \cong R/I_i$. Thus U/U_0 is a homomorphic image of $\sum_{i=1}^{n} \oplus R/I_i$. Since R satisfies the large condition, $|R/I_i| < \alpha$ for all i and hence $|U/U_0| < \alpha$. Thus U is critical.

(3). By (1) an α-critical C is nonsingular. Let $0 \neq x \in C$. Then x^r is not large. Hence R/x^r contains a submodule isomorphic to a right ideal of R.

A module with Krull dimension is called smooth if for every nonzero submodule N of M, $|N| = |M|$. It is straightforward to show that a semicritical module M with $\bigcap_{i=1}^{n} K_i = 0$ is smooth if and only if $|M/K_i| = |M|$ for all i. If we restrict our attention to smooth modules, we obtain a characterization for modules satisfying the large condition.

2.7 Proposition. Let M be a smooth right module with Krull dimension α and suppose M satisfies the large condition. If I is any complement submodule, then M/I is semicritical and smooth.

Proof: By [5,1.14] if I is a complement submodule, then $I = \bigcap_{i=1}^{n} I_i$ where the I_i are maximal complements, and hence are maximal nonlarge submodules. Since M is smooth, $|M/I_i| = \alpha$. If $I \supsetneq I_i$, then $I \leq_e M$ and by the large condition $|M/I| < |M| = |M/I_i|$. Thus M/I_i is α-critical.

2.8 Corollary. Let M be a smooth module with Krull dimension. Then M is semicritical if and only if M satisfies the large condition.

Proof: If M is semicritical by 2.4, M satisfies the large condition. The converse follows from 2.7 since 0 is a complement module.

As in 4.7 and 4.8 of [1], we have the following corollary which is true for semicritical rings of Krull dimension 1.

2.9 Corollary. Let R be a ring satisfying the large condition where $|R| = 1$.

(1) Every indecomposable injective right R-module is smooth.

(2) $\bigcap_{i=1}^{\infty} J^i = 0$, where J is the Jacobson radical of R.

3. k-primitive rings. An ideal D of a ring R is said to be k-coprimitive provided R/D has a finitely generated faithful critical right R-module C, where $|C| = k$. We shall say that an ideal is coprimitive if it is k-coprimitive for some k. A ring R is k-primitive if 0 is a k-coprimitive ideal. This definition represents a useful change from the definition in [3] where the faithful critical C was required to be cyclic. If the faithful critical for R/D is cyclic, we call D a cyclic coprimitive ideal.

The Krull radical K(R) of a ring R is the intersection of all cocritical right ideals. In [3] it was shown that K(R) is the intersection of the cyclic coprimitive ideals of R. We first show that the Krull radical is the intersection of the coprimitive ideals of R with the new definition.

3.1 Lemma. Let U be a uniform module and let $C_1,\ldots,C_n$ be critical submodules of U. Then $C = \sum_{i=1}^{n} C_i$ is critical and $|C| = |C_i|$ for all i.

Proof: Let $\alpha = |C_i|$ for all i, then directly $|\sum_{i=1}^{n} C_i| = \alpha$. Let H be a nonzero submodule of $\sum_{i=1}^{n} C_i \subset U$. Consider the chain $H \subset C_1 + H \subset C_1 + C_2 + H \subset \ldots \subset C_1 + C_2 + \ldots + C_n + H = \sum_{i=1}^{n} C_i$. Since U is uniform, $H \leq_e \sum_{i=1}^{n} C_i$, and $H \cap C_i \neq 0$ for all i. Then for $j \leq n$, $C_1 + \ldots + C_j + H/C_1 + \ldots + C_{j-1} + H \cong C_j/C_j \cap C_1 + \ldots + C_{j-1} + H$. Since C_j is α-critical, these factor modules have Krull dimension $< \alpha$. Since $|\sum_{i=1}^{n} C_i/H| = \sup_{j\leq n}\{|\sum_{i=1}^{j} C_i + H/\sum_{i=1}^{j-1} C_i + H|\}$, then $|\sum_{i=1}^{n} C_i/H| < \alpha$.

Note that an ideal D is cyclic k-coprimitive if and only if D is the largest ideal contained in a k-cocritical right ideal. We characterize the coprimitive ideals in a similar fashion with the following proposition.

3.2 Proposition. An ideal D of a ring R is k-coprimitive if and only if D is the largest ideal contained in $\bigcap_{i=1}^{n} K_i$, where the K_i are k-cocritical right ideals and the injective hulls $E(R/K_i)$ are isomorphic for all i.

Proof: Suppose R/D has a faithful finitely generated k-critical module C, say $C = \sum_{i=1}^{n} x_i R$. Then D is the largest ideal contained in $\bigcap_{i=1}^{n} x_i^r$. Furthermore, $E(R/x_i^r) \cong E(C)$ for all i.

Conversely, let D be the largest ideal contained in $\bigcap_{i=1}^{n} K_i$, where the K_i satisfy the hypothesis. Let $I = E(R/K_1)$. Since $E(R/K_i) \cong I$ for all i, there exists a monomorphism $\theta_i : R/K_i \to I$. Let $C = \sum_{i=1}^{n} \theta_i(R/K_i) \subset I$. Then C is a finitely generated R-module with annihilator D. Furthermore, by 3.1, C is critical and $|C| = k$. Thus R/D is a k-primitive ring.

From [3] and 3.2 we have the following result for the Krull radical.

3.3 Proposition. If D is a k-coprimitive ideal of a ring R, then D is a finite intersection of cyclic k-coprimitive ideals. Furthermore, the Krull radical equals the intersection of the coprimitive ideals of R.

We call a semicritical module M isotopic provided there exists a finite collection of cocritical submodules $K_1, \ldots, K_n$ such that $\bigcap_{i=1}^{n} K_i = 0$ and $E(M/K_i) \cong E(M/K_j)$ for all i and j. In this case, $|M/K_i| = |M/K_j|$ for all i and j. A semicritical ring R is isotopic if it is isotopic as a right R-module.

We have the following known results which relate these notions to prime and semiprime rings with Krull dimension.

3.4 Proposition. Let R be a ring with Krull dimension. Then

(1) R is semicompressible and isotopic if and only if R is a prime ring.

(2) R is semicompressible if and only if R is a semiprime ring.

We now show that k-primitive rings are related to semicritical modules in the same way that prime rings are related to semicompressible modules. In doing so we obtain a convenient characterization for k-primitive rings having a nonsingular finitely generated faithful critical. If the faithful critical is cyclic, these rings are called nonsingularly k-primitive in [2].

3.5 Theorem. The following are equivalent for a ring R.

(1) R is a semicritical isotopic ring and $|R| = \alpha$.

(2) R is α-primitive and $|R| = \alpha$.

(3) R has a faithful critical module C with $|C| = |R| = \alpha$.

(4) R has a faithful critical module C, where

$Z(C) = 0$ and $|C| = \alpha$.

Proof: (1) $\to$ (2). Since R is semicritical, we have that $0 = \bigcap_{i=1}^{n} K_i$, where R/K_i is critical for all i. Since $E(R/K_i) \cong E(R/K_j)$ for all i,j, then $|R/K_i| = |R/K_j| = \alpha$. Thus, as in 3.2, let $I = E(R/K_1)$. There exist monomorphisms $\theta_i\colon R/K_i \to I$. Thus $C = \sum_{i=1}^{n} \theta_i(R/K_i)$ is a finitely generated faithful right R-module and, by 3.1, is α-critical.

(2) $\to$ (3). This follows immediately from the definition.

(3) $\to$ (4). Let $0 \neq x_1 \in C$. Since C is faithful, $\bigcap_{x \in C} x^r = 0$ and, hence, if $x_1^r \neq 0$, there exists $x_2 \in C$ such that $x_1^r \neq x_1^r \cap x_2^r$. In this way, we form a proper chain $R \supset x_1^r \supset x_1^r \cap x_2^r \supset \ldots \supset \bigcap_{i=1}^{m} x_i^r \supset \ldots$, where the Krull dimension of the factors is $|C| = \alpha$. Since $|R| = |C| = \alpha$, it follows that the chain must terminate at zero. Consequently, there exists $x_1,\ldots,x_n \in C$ with

$\bigcap_{i=1}^{n} x_i^r = 0$. Thus R is semicritical and isotopic.

(1) and (3) $\to$ (4). Let $0 \neq x \in Z(C)$. Then $xR \cong R/L$ for some large right ideal L. By 2.4, R satisfies the large condition and hence $|C| = |xR| = |R/L| < |R| = \alpha$. This is impossible and, hence $Z(C) = 0$.

(4) $\to$ (3). Since $Z(C) = 0$, we immediately have that $Z(R) = 0$. Let X be the collection of annihilators of finite subsets of C. By [6,1.24], X satisfies the descending chain condition. Since C is faithful, $\bigcap_{x \in C} x^r = 0$ and there exists a finite subset $x_1,\ldots,x_n$ such that $\bigcap_{i=1}^{n} x_i^r = 0$. This implies the existence of an R-monomorphism $R \to \sum_{i=1}^{n} R/x_i^r \to C^{(n)}$. Thus $|R| \leq |C^{(n)}| = |C| \leq |R|$, and $|R| = |C| = \alpha$.

If R is semicompressible, then from 3.4, R is a subdirect product of prime rings. We now show that a similar result follows for semicritical rings in that they are a subdirect product of k-primitive rings.

3.6 Proposition. (1) If R is a semicritical ring, then R is a finite subdirect product of rings R_i, where R_i is k_i-primitive.

(2) R is a nonsingular semicritical ring if and only if R is a finite subdirect product of rings R_i, where R_i is k_i-primitive and $|R_i| = k_i$.

Proof: (1). If R is semicritical, there exists a finite set of cocritical right ideals K_i such that $\bigcap_{i=1}^{n} K_i = 0$. There exists an integer $m \leq n$ such that for a suitable arrangement of the subscripts $E(R/K_i) \ncong E(R/K_j)$ for $i \neq j$ and $1 \leq i, j \leq m$. Let $S_i = \{K_j \mid E(R/K_j) \cong E(R/K_i)\}$ for $1 \leq i \leq m$. Let D_i be the largest ideal in $\bigcap_{K_j \in S_i} K_j$. Then, by 3.2, D_i is coprimitive and clearly $\bigcap_{i=1}^{m} D_i = 0$.

(2). Since $Z(R) = 0$ and K_i is not large in R, then $Z(R/K_i) = 0$ over R and hence over R/D_i. This implies that R/D_i has a nonsingular finitely generated k_i-critical. Thus by 3.5, R/D_i is k_i-primitive with $|R/D_i| = k_i$.

The converse follows directly from 3.5.

4. Properties of k-primitive rings R, where $|R| = k$. In this section R will always be used to denote a k-primitive ring, where $|R| = k$, C will always be used to denote a finitely generated faithful k-critical of R and P will denote ass C, the assassinator of C. We now examine such a ring R and show that it has many properties which are similar to properties of prime rings with Krull dimension.

If Q is a prime ideal of a ring A, then A is said to be Q-primary if whenever $CB = 0$ and $C \neq 0$, then $B \subset Q$, where B and C are right ideals of R.

4.1 Proposition. The ring R satisfies the following.

(1) R is a semicritical isotopic ring and $Z(C) = 0$.

(2) R satisfies the large condition.

(3) R is smooth and $Z(R) = 0$.

(4) R is P-primary.

Proof: (1), (2) and (3) follow from 2.4 and 3.5. The proof of (4) is direct.

4.2 Proposition. The ideal P of R is the unique nonlarge minimal prime ideal of R, and $|R/P| = |R|$.

Proof: Since $Z(C) = 0$ and P annihilates a submodule of C, necessarily P is not large. From [1,3.1], then P is a minimal prime ideal. Further, since R is P-primary, if H is any ideal which is not large, then $H \subset P$. Consequently, P is the only nonlarge minimal prime.

Finally, since P is not large and R is smooth, then $|R/P| = |R|$.

A k-primitive ring with Krull dimension k can have many distinct finitely generated faithful k-critical modules. The relationship between these faithful modules is given in the following proposition.

4.3 Proposition. The following hold for the ring R.

(1) Every uniform right ideal of R is subisomorphic to C.

(2) Every k-critical R-module has assassinator P.

(3) If D is a faithful k-critical module for R, then $E(D) \cong E(C)$ and ass D = ass C = P.

Proof: (1). If U is a nonzero uniform right ideal of R, then $CU \neq 0$. Hence there exists $x \in C$ such that $xU \neq 0$. Since $Z(C) = 0$, $U \cong xU \subset C$.

(2). Let G be a k-critical module. By 4.1 and 2.6, $Z(G) = 0$. Since P = ass C is the only nonlarge prime, necessarily P = ass G.

(3). This follows directly from (1) and (2).

4.4 Lemma. Let Q be a prime ideal of a ring A with Krull dimension. If $|A| = |A/Q|$, then for every critical right ideal D of R, either $D \cap Q = 0$ or $D \subseteq Q$.

Proof: Suppose $D \cap Q \neq 0$ and $D \not\subset Q$. Then $0 \neq D/D \cap Q \cong D + Q/Q \subset A/Q$. Since A/Q is smooth, $|D + Q/Q| = |A/Q| = |A|$. However, D is critical which implies that $|D/D \cap Q| < |D| \leq |A|$. This is impossible.

If A is a prime ring with Krull dimension then the compressible right ideals of A are subisomorphic. This result remains true for k-primitive rings with Krull dimension k.

4.5 Proposition. If U is a uniform right ideal of R and $U \not\subset P$, then U is compressible. In addition, the compressible right ideals of R are subisomorphic.

Proof: Let U be a uniform right ideal of R such that $U \not\subset P$. Then by 4.4, $U \cap P = 0$. Thus U is isomorphic to a uniform right ideal of R/P, and therefore U is compressible.

Since P is not large there exists a compressible right ideal U such that $U \cap P = 0$. Since R is P-primary, if K is any other compressible right ideal, then $KU \neq 0$. Thus there exists $x \in K$ with $xU \neq 0$ and a monomorphism $U \to xU \subset K$. Since K is compressible, then K is subisomorphic to xU, and hence to U. Thus U and any other compressible right ideal are subisomorphic. Since being subisomorphic is a transitive property, any two compressible right ideals are subisomorphic.

4.6 Proposition. If H is the center of R, then the nonzero elements of H are regular.

Proof: As in [3], H can be embedded into $\mathrm{End}_R C$, and thus $H \cap P = 0$. If $a \in H$ and $a^r \neq 0$, then $a^r \cdot aR = 0$ and since R is P-primary, $a \in P$. Thus $a = 0$.

4.7 Proposition. Let I be a proper closed right ideal. Then

(1) R/I is semicritical.

(2) If D is the largest ideal contained in I, then R/D is k-primitive, $|R/D| = k$, and the faithful k-critical for R/D is $C^* \subset E(C)$. Furthermore, $D \subset P$ and ass $C^* = P/D$.

Proof: (1). By 4.1, $Z(R) = 0$. Thus if I is closed, I is a complement right ideal. By 2.7, R/I is semicritical.

(2). Since I is a complement, $I = \bigcap_{i=1}^{n} I_i$ implies I_i is not large for all i. Hence by 4.3, R/I_i contains a submodule which is isomorphic to a submodule of C. Thus $E(R/I_i) \cong E(C)$ for all i. Consequently, by 3.2, D is k-coprimitive. Further, the proof of 3.2 shows that since $E(R/I_i) \cong E(C)$ for all i, E(C) contains a finitely generated k-critical C* with $(C^*_R)^r = D$. Finally since $(C \cap C^*)D = 0$, then $D \subset$ ass C. Since ass C = P, then ass C* = P/D over R/D.

4.8 Proposition. Let $P_1,\ldots,P_n$ be the set of minimal primes of R with P_1 = ass C and $N = \bigcap_{i=1}^{n} P_i$, the prime radical. Then

(1) $|P_1/N| < |R|$.

(2) $|P_i/N| = |R|$, $2 \leq i \leq n$.

(3) If R/N is smooth, then N is a prime ideal.

Proof: (1). By 4.2, $P_2,\ldots,P_n$ are large and since by 4.1, R satisfies the large condition, $|R/P_2 \cap \ldots \cap P_n| < |R|$. Thus $P_1/N = P_1 + P_2 \cap \ldots \cap P_n / P_2 \cap \ldots \cap P_n$ implies that $|P_1/N| < |R|$.

(2). Since P_i is large for $2 \leq i \leq n$, $|R/P_i| < |R|$. Further $|R/N| = |R|$ and $|R/N| = \sup\{|R/P_i|, |P_i/N|\}$. Necessarily then $|P_i/N| = |R|$.

(3). Now $|R/N| = |R|$ and $|P_1/N| < |R|$. If R/N is smooth, then $P_1/N = 0$ and $N = P_1$.

5. k-primitive rings and artinian classical quotient rings. Piecewise domains, which are k-primitive rings with Krull dimension k, are discussed in [2]. These rings, like prime rings with Krull dimension, have an artinian classical quotient ring. It is not known if this is true for all k-primitive rings with Krull dimension k. In this section we shall discuss the properties of the artinian classical quotient ring of such rings when it does exist.

In [4] it is shown that for a prime ring with Krull dimension, the minimal right ideals of the classical right quotient ring Q are of the form UQ where

U is compressible right ideal of R. The same result is true for k-primitive rings R, where $|R| = k$, when they have an artinian classical quotient ring.

5.1 Proposition. Let R be a k-primitive ring with $|R| = k$ and suppose that R has an artinian classical quotient ring Q. Then D is a minimal right ideal of Q if and only if D is of the form UQ where U is a compressible right ideal of R. Furthermore, all the minimal right ideals of Q are isomorphic.

Proof: Let C denote a faithful k-critical for R and P = ass C. By 4.5, there exists a compressible right ideal U such that $U \cap P = 0$. Let $U_0 = UQ \cap R$. Then $U_0 \cap P = 0$ and by 4.5, U_0 is also compressible. This implies that U_0Q is a compressible right ideal of Q. For if K is a nonzero right ideal of Q contained in U_0Q, then $0 \neq K \cap R \subset U_0Q \cap R = U_0$ and thus there exists a monomorphism $\theta: U_0 \to K \cap R$. By [10,1.6], θ extends to a monomorphism $U_0Q \to K$. Thus U_0Q is a compressible right ideal in an artinian ring and consequently U_0Q, and hence UQ, is a minimal right ideal of Q.

Now suppose C_0 is any compressible right ideal of R. By 4.5, there exists a monomorphism of $C_0 \to U$. This extends to a monomorphism $\theta^*: C_0Q \to UQ$. Thus C_0Q is a minimal right ideal of Q.

Conversely, if D is a minimal right ideal of Q then $D \cap R$ contains a nonzero compressible right ideal C_0 since $Z(R) = 0$. Thus $C_0Q = D$.

Finally note that the above arguments show that every minimal right ideal of Q is isomorphic to UQ.

5.2 Theorem. Let R be a k-primitive ring with $|R| = k$ and let C be a faithful k-critical R module, with P = ass C. If R has an artinian classical quotient ring Q, the following hold.

(1) If U is a compressible right ideal of R such that $U \cap P = 0$, then $End_R U = End_{R/P} U$ has an artinian classical quotient ring $D = End_Q UQ = End_K UQ$, where $K = Q/PQ$.

(2) $Q = eQ \oplus PQ$ as right ideals, where eQ is a ring which is isomorphic to the classical quotient ring Q(R/P) of R/P. Here $Q(R/P) \cong D_m$, where D is the division ring of (1).

(3) $Q \cong \begin{bmatrix} (1-e) & Q(1-e) & (1-e)Qe \\ & 0 & eQ \end{bmatrix}$, where

$eQ \cong D_m$ and $PQ = \begin{bmatrix} (1-e)Q(1-e) & (1-e)Qe \\ 0 & 0 \end{bmatrix}$.

Proof: (1). Since U is isomorphic to a right ideal of R/P, the result follows from [4, theorem 4].

(2). Let S be the sum of all the compressible right ideals of R. Then $S + P/P$ is a large ideal in R/P and hence $S + P/N$ is large in R/N. Thus $S + P/N$ contains a regular element $\bar{r}$ of R/N and since by [8], R satisfies the regularity condition, $r \in S + P$ is regular in R. Then $r \in \sum_{i=1}^{n} C_i + P$ where C_i is compressible and $C_i \cap P = 0$. Thus $Q = (\sum_{i=1}^{n} C_i + P)Q \subseteq \sum_{i=1}^{n} C_iQ + PQ \subseteq Q$ where C_iQ is a minimal right ideal of Q by 5.1, and $C_iQ \cap PQ = 0$ for all i. Since the C_iQ are minimal right ideals, we can form the direct sum $Q = \sum_{i=1}^{m} \oplus C_iQ \oplus PQ$. Hence Q/PQ is ring isomorphic to $\sum_{i=1}^{m} \oplus C_iQ$ which is isomorphic to D_m by (1).

(3). This follows directly from the Peirce decomposition since PQ is a two-sided ideal.

One can prove directly that certain properties of a k-primitive ring R with $|R| = k$ carry over to its classical quotient ring. Here $CQ = C \otimes Q$ as in 1.5 of [10].

5.3 Proposition. Let R be a k-primitive ring with $|R| = k$, faithful k-critical C and P = ass C. If R has an artinian classical quotient ring Q, then CQ is a faithful uniform Q-module, ass CQ = PQ, Q is PQ-primary, and $Z(CQ) = 0$.

In [2] several questions regarding the structure of k-primitive rings were raised. If N(R) is prime, is R necessarily prime? Does R always have an artinian classical quotient ring? A negative answer to the first implies a negative answer to the second.

5.4 Proposition. Let R be a k-primitive ring with $|R| = k$. If N(R) is a prime ideal in R, then R does not satisfy the regularity condition.

Proof: Since N = N(R) is not large, there exists a right ideal $I \neq 0$ such that $I \cap N = 0$. This implies that the two-sided ideal $N^{\ell} = \{r \varepsilon R | rN = 0\}$ has members not in N. Since N is prime, $N^{\ell} + N/N$ is large in R/N and so contains an element a + N with $a \varepsilon N^{\ell}$ which is regular in R/N. Clearly a cannot be regular in R.

Let R be a k-primitive ring where $|R| = k$. The example in section 4 of [2] shows that in general the right artinian classical quotient ring for R is not 0-primitive. We conclude this section with the following question. Can R be embedded in a localization R_T, such that R_T is α-primitive where $|R_T| = \alpha$, and α is strictly less than k for $k > 1$?

References

1. A. K. Boyle and E. H. Feller, Smooth Noetherian Modules, Comm. in Alg. 4 (1976), 617-637.

2. A. K. Boyle, M. G. Deshpande, and E. H. Feller, On Nonsingularly k-primitive rings, Pac. J. Math. 68 (1977), 303-311.

3. M. G. Deshpande and E. H. Feller, The Krull radical, Comm. in Alg. 4 (1975), 185-193.

4. A. W. Goldie, The Structure of Prime Rings under the Ascending Chain Condition, Proc. London Math. Soc. 8 (1958), 589-698.

5. ___________, The Structure of Noetherian Rings, in Lecture Notes in Mathematics, No. 246, Springer-Verlag (1972), 213-321.

6. K. R. Goodearl, Singular torsion and the splitting properties, Amer. Math. Soc. Memoirs, No. 124 (1972).

7. R. Gordon and J. C. Robson, Krull dimension, Amer. Math. Soc. Memoirs, No. 133 (1973).

8. R. Gordon, Gabriel and Krull dimension, in Ring Theory, Lecture Notes in Pure and Applied Math (7), Marcel Dekker, 1974.

9. G. Krause, Descending chains of submodules and the Krull dimension of Noetherian modules, J. of Pure and Applied Alg. 3 (1973), 385-394.

10. L. Levy, Torsion free and divisible modules, Can. J. Math, 15 (1963), 132-151.

A NOTE ON LOEWY RINGS AND CHAIN CONDITIONS ON PRIMITIVE IDEALS

V. P. Camillo and K. R. Fuller*
The University of Iowa
Iowa City, Iowa 52242, U.S.A.

All rings are associative with identity. The socle series for a module M is defined transfinitely by $\mathrm{Soc}_0(M) = 0$, $\mathrm{Soc}_{\alpha+1}(M)/\mathrm{Soc}_\alpha(M) = \mathrm{Soc}(M/\mathrm{Soc}_\alpha(M))$ and, if α is a limit ordinal, $\mathrm{Soc}_\alpha(M) = \cup_{\beta<\alpha} \mathrm{Soc}_\beta(M)$ (see [2, p. 470]). If $M = \mathrm{Soc}_\alpha(M)$, M is called a *Loewy module* [3], [4] and its *Loewy length* is the smallest such ordinal α. A ring R is called a *right Loewy ring* (or said to be right semi-artinian) in case R_R is a Loewy module or, equivalently, every nonzero right R-module contains a minimal submodule.

In [3, Remark (3.1)] we observed that the results of [2], [7], [9], [3] show that any right Loewy ring that is semilocal (hence left perfect), commutative, or of finite Loewy length has the property that each of its nonzero left modules has a maximal submodule. These three kinds of rings satisfy the ascending chain condition (*a.c.c.*) on primitive ideals. Here we prove that over any right Loewy ring with a.c.c. on primitive ideals every nonzero left module has a maximal submodule, and we provide an example of a Loewy ring with a module ($\neq 0$) containing no maximal submodules. Since Kaplansky [6] proved primitive factor rings of a ring satisfying a polynomial identity (i.e., a P.I. ring) are simple artinian, it follows that the nonzero left modules over a right Loewy P.I. ring all have maximal submodules.

If ${}_R M$ is a left R-module with maximal submodule $L < M$ and P is the left annihilator $P = \ell_R(M/L)$ then P is a left primitive ideal and $PM \neq M$. If R is right Loewy, this version of Nakayama's lemma holds whether M has a maximal submodule or not. Before proving so, we recall that the left and right socles of a left or right primitive ring with minimal right ideals (a so-called PMI ring) are equal and faithful (see [5, Chapter IV]). In particular, over a right Loewy ring left primitive and right primitive ideals are one and the same.

1. LEMMA: Let M be a nonzero left module over a right Loewy ring R. Then there is a primitive ideal P in R such that $PM \neq M$.

Proof. Let $A = \ell_R(M)$, let $A \leq T_R \leq R_R$ such that T/A is a minimal right ideal of R/A, and let $P = r_R(T/A)$. Then

$$TPM \subseteq AM = 0 \neq TM$$

and so $PM \neq M$.

This lemma is the key[1] to our

2. THEOREM. *If R is a right Loewy ring with* a.c.c. *on (left or right) primitive ideals then every nonzero left R-module has a maximal submodule.*

Proof. Let ${}_RM \neq 0$. Then by the lemma there is a primitive ideal P maximal among those satisfying $PM \neq M$. Let $S/P = \mathrm{Soc}(R/P)$. If $SM \neq M$ then, again by the lemma, there is a primitive ideal Q/S in R/S such that $(Q/S)\cdot(M/SM) \neq M/SM$. But then Q is primitive in R, Q properly contains P and $QM \neq M$, contrary to the maximality of P. Thus $SM = M$, and we see that

$$M/PM = SM/PM = S/P \cdot M/PM$$

is a nonzero semisimple factor module of M. So, as promised, M does have a maximal submodule.

Năstăsescu and Popescu [9] proved that if R is a commutative Loewy ring then R/Rad R is von Neumann regular. It follows [9, Remark 1, p. 361] and [7, Proposition 1.6] that every R-module has a maximal submodule. Although we do not know whether Loewy P.I. rings are von Neumann regular modulo the radical (but see [1, Theorem 2]), in the presence of Kaplansky's theorem [6], our theorem immediately yields the following result.

2. COROLLARY. *Let R satisfy a polynomial identity. If R is right Loewy then every nonzero left R-module has a maximal submodule.*

Some condition on a Loewy ring R is necessary for this conclusion to hold. (Perhaps a.c.c. on primitive ideals.)

4. EXAMPLE: *A left and right Loewy ring over which not every nonzero module has a maximal submodule.*

Demonstration: Let K be a field and let V be a countably infinite dimensional K-vector space

$$V = \oplus_{i=1}^{\infty} Kv_i \qquad (v_i \in V).$$

Let $f_i : \mathbb{N} \longrightarrow \mathbb{N}$ $(i \in \mathbb{N} = \{1,2,\cdots\})$ be a sequence of injections whose images form a partition of $\mathbb{N}$, so that

$$\mathbb{N} = \dot{\bigcup}_{\mathbb{N}} f_i(\mathbb{N}),$$

and write

$$f_i(n) = i_n \qquad (i,n \in \mathbb{N}).$$

The ring we shall construct is a subring of $\mathrm{End}({}_K V)$ (with maps written on the right). In the construction, if $V = V_1 \oplus V_2$, when we write $t : V_1 \longrightarrow V_j$ we also mean $t : V_2 \longrightarrow 0$

and $t \in \mathrm{End}({}_K V)$. First we define decompositions $V = \oplus_{i\in\mathbb{N}} V_{ix}$ $(x \in \mathbb{N})$: Let

$$V = \oplus_{i\in\mathbb{N}} V_{i1} \quad \text{with} \quad V_{i1} = Kv_i$$

and, given $V = \oplus_{i\in\mathbb{N}} V_{ix}$, let

$$V = \oplus_{i\in\mathbb{N}} V_{ix+1} \quad \text{with} \quad V_{ix+1} = \oplus_{n\in\mathbb{N}} V_{i_n x}.$$

Then we define linear transformations of V, $e_{ijx} : V_{ix} \longrightarrow V_{jx}$ $(i,j,x \in \mathbb{N})$: Let

$$e_{ij1} : cv_i \longmapsto cv_j \qquad (c\varepsilon K)$$

and, given $e_{ijx} : V_{ix} \longrightarrow V_{jx}$ $(i,j \in \mathbb{N})$, let

$$e_{ijx+1} = \oplus_{\mathbb{N}} e_{i_n j_n x} : V_{ix+1} \longrightarrow V_{jx+1}.$$

Now, observing that any (finite) linear combination of the e_{ijx} with $x < y$ annihilates all but finitely many of the V_{iy-1}, but that no nontrivial linear combination of the e_{ijy} does so, we see that

$$\{1_V\} \cup \{e_{ijx} \mid i,j,x \in \mathbb{N}\}$$

is a K-linearly independent set of maps in $\mathrm{End}({}_K V)$. Let R be the subspace of $\mathrm{End}({}_K V)$ that this set spans, and let

$$S_0 = 0$$

and, for each $x > 0$,

$$S_x = \Sigma_{\substack{w \le x \\ i,j \in \mathbb{N}}} Ke_{ijw}.$$

Then we have equations

(R1) $$e_{ijx} e_{k\ell x} = \begin{cases} 0 , & \text{if } j \ne k \\ e_{i\ell x} , & \text{if } j = k, \end{cases}$$

(R2) $$e_{ijx} e_{k\ell x+1} = \begin{cases} 0 , & \text{if } j \notin \{k_n \mid n \in \mathbb{N}\} \\ e_{i\ell_n x} , & \text{if } j = k_n \end{cases}$$

(R3) $$e_{ijx+1} e_{k\ell x} = \begin{cases} 0 , & \text{if } k \notin \{j_n \mid n \in \mathbb{N}\} \\ e_{i_n \ell x} , & \text{if } k = j_n \end{cases}$$

which show that

(R4) $$S_x S_y = S_{\min\{x,y\}}.$$

It follows that R is a subalgebra of $\mathrm{End}({}_K V)$ with ideals $0 = S_0 < S_1 < \cdots < \bigcup_{\mathbb{N}} S_x < R$; and by the Structure Theorem for

PMI Rings [5, p. 75], R is a primitive algebra with $\mathrm{Soc}(R) = S_1$. Now checking that the K-linear map Φ determined by

$$\Phi : e_{ijx} \longmapsto \begin{cases} e_{ijx-1}, & x > 1 \\ 0, & x = 1, \end{cases}$$

$$\Phi : 1_V \longmapsto 1_V$$

is a surjective ring homomorphism

$$\Phi : R \longrightarrow R$$

such that

$$\Phi(S_x) = S_{x-1} \qquad x \in \mathbb{N},$$

we see that each S_x is primitive,

$$\mathrm{Soc}_x({}_RR) = S_x = \mathrm{Soc}_x(R_R),$$

and R is a left and right Loewy ring of length $\omega+1$.

We use similar methods to construct a submodule $M \neq 0$ of the left R-module $V^* = \mathrm{Hom}_K({}_KV_R, K)$ such that M has no maximal submodule. Define linear functionals on V $\gamma_{ix} : V_{ix} \longrightarrow K$ $(i, x \in \mathbb{N})$ by

$$\gamma_{i1} : cv_i \longmapsto c \qquad (c \in K)$$

and, given $\gamma_{ix} : V_{ix} \longrightarrow K \quad (i \in \mathbb{N})$,

$$\gamma_{ix+1} = \oplus_{n \in \mathbb{N}} \gamma_{i_n x} : V_{ix+1} \longrightarrow K.$$

Upon observing that the functionals

$$\{\gamma_{ix} \mid i,x \in \mathbb{N}\}$$

are independent, we let M be the subspace of V^* that they span, and write

$$L_0 = 0$$

and

$$L_x = \sum_{\substack{w \leq x \\ i \in \mathbb{N}}} K\gamma_{ix}.$$

Here we have equations

$$\text{(M1)} \qquad e_{ijx}\gamma_{kx} = \begin{cases} 0, & \text{if } j \neq k \\ \gamma_{ix}, & \text{if } j = k \end{cases}$$

$$\text{(M2)} \qquad e_{ijx}\gamma_{kx+1} = \begin{cases} 0, & \text{if } j \notin \{k_n \mid n \in \mathbb{N}\} \\ \gamma_{ix}, & \text{if } j = k_n \end{cases}$$

(M3) $$e_{ijx+1}\gamma_{kx} = \begin{cases} 0 & , \text{ if } k \notin \{j_n \mid n \in \mathbb{N}\} \\ \gamma_{i_n x} , & \text{if } k = j_n . \end{cases}$$

From these and (R4) it follows that

(M4) $$S_x L_y = L_{\min\{x,y\}}$$

so that M is a left R-submodule of V^* with submodules

$$0 = L_0 < L_1 < \cdots < \bigcup_{\mathbb{N}} L_x = M.$$

Using (M1) it is easy to check that L_{x+1}/L_x is generated by each of its nonzero elements, and hence is simple. On the other hand, if $y \in \mathbb{N}$ and

$$\gamma \in L_{y+1} \setminus L_y$$

then we must have

$$\gamma = \sum a_i \gamma_{iy+1} + \gamma' \quad \text{with} \quad \gamma' \in L_y$$

and some $a_k \neq 0$. Since $V_{k_n y}\gamma' = 0$ for all but finitely many $n \in \mathbb{N}$, there is an n with $e_{1k_n y}\gamma' = 0$. Thus by (M2) we have

$$\begin{aligned} e_{1k_n y}\gamma &= \sum a_i e_{1k_n y}\gamma_{iy+1} + e_{1k_n y}\gamma' \\ &= a_k \gamma_{1y} \in L_y \setminus L_{y-1} . \end{aligned}$$

This proves that if $x < y$ then L_y/L_x is essential in L_{y+1}/L_x and hence, by induction, L_{x+1}/L_x is essential in $\bigcup_{y \geq x} L_y/L_x = M/L_x$. Now it follows that the L_x are the only submodules of M, so M has no maximal submodule.

We note that the ring of our example is von Neumann regular [8, Theorem 2]. Also its ideals are just the members of its Loewy series, so the ring has d.c.c. on ideals—a fact that suggests our concluding result.

5. PROPOSITION: <u>Every right Loewy ring has</u> d.c.c. <u>on primitive ideals</u>.

<u>Proof</u>. Suppose that $P_1 \supseteq P_2 \supseteq \cdots$ is a chain of primitive ideals in a right Loewy ring R. Assume, as we may, that $\bigcap_1^\infty P_i = 0$. Let T be a minimal right ideal in R. Then there is an n such that $T \cap P_n = 0$, so that

$$T \cong T + P_k/P_k \leq R/P_k$$

must be the faithful simple right R/P_k module for all $k \geq n$. But

$$TP_n \subseteq T \cap P_n = 0$$

so the faithfulness of T over R/P_k implies $P_n = P_k$ for all $k \geq n$.

REFERENCES

[1] E.P. Armendariz and J.W. Fisher, Regular P.I.-rings, Proc. Amer. Math. Soc. 39(1973), 247-251.

[2] H. Bass, Finitistic dimension and a homological generalization of semiprimary rings, Trans. Amer. Math. Soc. 95(1960), 466-488.

[3] V.P. Camillo and K.R. Fuller, On Loewy length of rings, Pacific J. Math. 53(1974), 347-354.

[4] L. Fuchs, Torsion preradicals and ascending Loewy series of modules, J. Reine Angew. Math. 239/240(1970), 169-179.

[5] N. Jacobson, Structure of Rings, Amer. Math. Soc. Coll. Pub. (1964), Providence, R.I.

[6] I. Kaplansky, Rings with a polynomial identity, Bull. Amer. Math. Soc. 54(1948), 575-580.

[7] L.A. Koifman, Rings over which every module has a maximal submodule, Math Notes of the Acad. of Sci. of the USSR, vol. 7, no. 3, Mar.-April 1970.

[8] C. Năstăsescu, Quelques remarques sur une classe d'anneaux, C.R. Acad. Sci. Paris 270(1 Avril 1970).

[9] C. Năstăsescu and N. Popescu, Anneaux Semi-artinians, Bull. Soc. Math. France. 96(1968), 357-368.

FOOTNOTES

* Fuller's research was partially supported by N.S.F. Grant MCS77-00431.

1 It also follows from the lemma that if R is a right Loewy ring then the left R-modules all have maximal submodules iff the left modules over the primitive factor rings of R all do.

DECOMPOSITION OF DUAL-CONTINUOUS MODULES

Saad Mohamed and Bruno J. Müller
Kuwait University, Kuwait
and
McMaster University, Hamilton, Ontario, Canada

Introduction. A module M is called d-continuous if (I) for every submodule A of M there exists a decomposition $M=M_1 \oplus M_2$ such that $M_1 \subset A$ and $M_2 \cap A$ is small in M_2, and (II) every surjection from M onto a summand of M splits.

We remind the reader that this definition is dual to that of a continuous module (Utumi [6], Mohamed and Bouhy [4]) which in turn extends that of a (quasi-)injective module, and that it generalizes the concept of a (quasi-)projective module if and only if the ring is perfect. Over an arbitrary ring, the relationship between d-continuity and (quasi-)projectivity is less close, but d-continuous modules still possess many properties which are dual to those of continuous, in particular (quasi-)injective modules.

Mohamed and Singh ([5], Theorem 4.7) obtained this decomposition theorem: Let M be a d-continuous module. Then $M=N+N'$, where N' is a summand of M with Rad $N' = N'$, and $N = \Sigma\oplus_{i\in I} A_i$ where each A_i is local and the sum of finitely many of the A_i is always a summand of M. We improve this result as follows:

THEOREM 1. A d-continuous module M has a decomposition, unique up to isomorphism, $M = \Sigma\oplus_{i\in I} A_i \oplus N$, where each A_i is a local *module and* N = Rad N .

While summands of d-continuous modules are always d-continuous ([5], Proposition 3.5), the sum of d-continuous modules need not share this property; in fact d-continuity of M⊕M forces M to be quasi-projective ([5], Corollary 4.2). Our second result contributes to this question:

THEOREM 2 . *A module* M *is finitely generated and d-continuous if and only if* $M = \Sigma\oplus_{i=1}^{n} A_i$, *where each* A_i *is local d-continuous and* A_j*-projective for all* $j \neq i$.

Preliminaries. All our modules will be right-modules over a not necessarily commutative ring with identity element. Referring to modules, sum and summand will mean direct sum, summand. Rad M is the Jacobson radical of the module M , ie. the intersection of all maximal submodules; in particular Rad M = M if and only if M has no maximal submodules. A module M is called *local* if it has a proper submodule N which contains all proper submodules; then N = Rad M and M is cyclic, indeed generated by every $x \in M-N$.

A submodule A of M is called a *d-complement* (of B in M) if A is minimal with A + B = M . We write $A \subset_s M$ if A is small in M , ie. if A + B = M implies B = M .

For convenient reference we list a few known results:

(A) A is a d-complement of B in M if and only if A + B = M and

$A \cap B \subset_s A$ (Miyashita [3]) .

(B) A d-continuous modules is perfect in the sense of Miyashita (ie. for every pair of submodules with $A+B=M$, A contains a d-complement of B), and every d-complement submodule of M is a summand ([5], Proposition 3.7) .

(C) If B is a summand of the d-continuous module M , and if A is a d-complement of B in M , the $M=A \oplus B$ (ibid. Corollary 3.9) .

(D) A d-continuous module is indecomposable if and only if its endomorphism ring is local (ibid. Corollary 3.11) .

(E) If M is a direct sum of indecomposable modules, $M = \Sigma\oplus_{i \in I} M_i$, where each M_i is countably generated and has local endomorphism ring, then any other direct sum decomposition of M refines to a decomposition isomorphic to this one, and (in particular) any summand of M is again a direct sum of modules, each isomorphic to one of the original summands M_i (Warfield [7], Theorem 1) .

(F) Let M be finitely generated, or let I be finite. If M is M_i-projective for all $i \in I$, then M is $\oplus_{i \in I} M_i$-projective (Azumaya [1], Theorem 5 and Proposition 5; partly restated in [2], Proposition 1.16) .

Auxiliary Results. LEMMA 3 . *The following are equivalent for a d-continuous module* M :

(1) Rad M *is small in* M ;

(2) *every proper submodule of* M *is contained in a maximal submodule* ;

(3) M *is the sum of local modules.*

<u>Proof</u>. (2) implies (1) trivially, and (1) implies (3) via the cited Theorem 4.7 of [5] . Assume (3) , ie. $M = \Sigma\oplus_{i\in I} A_i$ where each A_i is local, and consider a proper submodule B of M . By d-continuity, $M=M_1\oplus M_2$ where $M_1\subset B$ and $M_2\cap B\subset_s M_2$; hence $B=M_1\oplus(M_2\cap B)$ and $M_2\cap B\subset_s M$. All A_i , being d-continuous, have local endomorphism rings by (D) , and therefore by (E) $M_2 = \Sigma\oplus_{k\in K} C_k$ where the C_k are isomorphic to some of the A_i . K is non-empty as B is proper, and with a fixed $\ell\in K$ and $D = M_1 \oplus \Sigma\oplus_{k\in K,k\neq\ell} C_k$, $M/D\oplus\text{Rad } C_\ell \simeq C_\ell/\text{Rad } C_\ell$ is simple, hence $D\oplus\text{Rad } C_\ell$ is maximal in M . But $D\oplus\text{Rad } C_\ell \supset M_1+\text{Rad } M \supset M_1\oplus(M_2\cap B) = B$; ie. (2) .

LEMMA 4 . <u>Let</u> $M = \Sigma\oplus_{i\in I} A_i$, <u>where each</u> A_i <u>is local and</u> A_j<u>-projective for all</u> $j\neq i$. <u>If</u> Rad M <u>is small, then every non-small submodule of</u> M <u>contains a nonzero summand.</u>

<u>Proof</u>. Let B be a non-small submodule of M ; then certainly $B \not\subset \text{Rad } M$ hence $\overline{B}\neq 0$ in $\overline{M}=M/\text{Rad } M = \Sigma_{i\in I}\overline{A}_i$, which is semisimple since $\overline{A}_i \simeq A_i/\text{Rad } A_i$ is simple. Therefore by an exchange argument, $\overline{M} = \overline{B} + \Sigma_{i\neq k}\overline{A}_i$ for some k , hence $M = B + \Sigma\oplus_{i\neq k}A_i + \text{Rad } M$, hence $M = B + \Sigma\oplus_{i\neq k}A_i$ since Rad M is small.

By (F) A_k is $\Sigma\oplus_{i\neq k}A_i$-projective, and consequently there exists a homomorphism Φ making the following diagram commute:

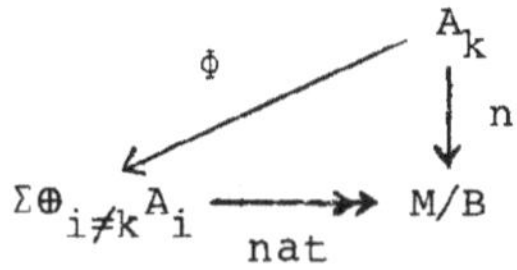

This means $a-\Phi(a)\in B$ for all $a\in A_k$, hence $A_k' = \{a-\Phi(a): a\in A_k\}$ is a nonzero submodule of B ; and one readily checks $M = A_k' \oplus \Sigma\oplus_{i\neq k}A_i$, ie. that A_k' is a summand of M .

LEMMA 5 . *Let A be a submodule of the d-continuous module M , let B_1 and B_2 be d-complements of A in M , and let A_o be a d-complement of B_1 in M contained in A . Then $M = A_o \oplus B_1 = A_o \oplus B_2$.*

Proof. By (B) , B_1 and B_2 are summands of M , and by (A) , $A \cap B_1 \subset_s B_1$ and $A \cap B_2 \subset_s B_2$. Since A_o is a d-complement of the summand B_1 , $M = A_o \oplus B_1$ by (C) . Then $A = A_o \oplus (A \cap B_1)$ and therefore $M = A + B_2 = A_o + (A \cap B_1) + B_2$; but $A \cap B_1 \subset_s B_1$ hence $\subset_s M$ and consequently $M = A_o + B_2$. Since $A_o \cap B_2 \subset A \cap B_2 \subset_s B_2$, (A) shows that B_2 is a d-complement of A_o in M , and therefore $M = A_o \oplus B_2$ by (C) .

COROLLARY 6 . *In a d-continuous module, d-complements are unique up to isomorphism.*

Proof. Combine Lemma 5 with (B) .

Proof of Theorem 1 . In the d-continuous module M , let A be a d-complement of Rad M , and let N be a d-complement of A contained in Rad M ; both exist and are summands, by (B) . Then $M = A \oplus N$ by (C) , and $A \cap \text{Rad } M \subset_s A$ by (A) ; therefore $\text{Rad } A = A \cap \text{Rad } M \subset_s A$ and $\text{Rad } N = N \cap \text{Rad } M = N$. Since A is d-continuous, $A = \Sigma\oplus_{i \in I} A_i$ where each A_i is local, by Lemma 3 , and consequently $M = \Sigma\oplus_{i \in I} A_i \oplus N$.

It remains to prove the uniqueness of this decomposition. Suppose $M = B\oplus N'$, where $B = \Sigma\oplus_{j\in J} B_j$ with all B_j local, and Rad $N' = N'$. By Lemma 3 applied to B , $B\cap \text{Rad } M = \text{Rad } B \subset_s B$. Moreover $B+\text{Rad } M \supset B\oplus N' = M$, hence B is a d-complement of Rad M by (A) , and therefore $M = A\oplus N = B\oplus N = A\oplus N' = B\oplus N'$ by Lemma 5 (since A and B are d-complements of Rad M , and N respectively N' are d-complements of A respectively B contained in Rad M) . Consequently $A\simeq B$ and $N\simeq N'$, and uniqueness follows from (D) and (E) .

<u>Proof of Theorem 2</u> . (1) Assume M to be finitely generated and d-continuous. Then $M = \Sigma\oplus_{i=1}^{n} A_i$ where each A_i is local (and d-continuous) by Lemma 3 , since Rad $M \subset_s M$ due to finitely generated-ness. That the A_i are A_j-projective for $j\neq i$, follows immediately from Proposition 4.1 of [5] .

(2) Assume conversely $M = \Sigma\oplus_{i=1}^{n} A_i$ with the properties described in the theroem. To verify condition (I) of the definition of d-continuity, consider a proper submodule B of M . Among all the decompositions $M = \Sigma\oplus B_j$ into indecomposable summands (which are all isomorphic by (E)) pick one with a maximal number t of summands contained in B , say $B_1\oplus\ldots\oplus B_t \subset B$.

Suppose $B_{t+1}\oplus\ldots\oplus B_n\cap B$ is not small in $B_{t+1}\oplus\ldots\oplus B_n$. Then , Lemma 4 applied to this module, produces a nonzero summand contained in $B_{t+1}\oplus\ldots\oplus B_n\cap B$, and a refinement according to (E) yields an isomorphic decomposition $B_{t+1}\oplus\ldots\oplus B_n = C_{t+1}\oplus\ldots\oplus C_n$ with $C_{t+1}\subset B$, hence $B_1\oplus\ldots\oplus B_t\oplus C_{t+1} \subset B$, contradicting the maximality of t .

This confirms condition (I) .

(3) To verify condition (II) we observe first the elementary fact that every homomorphism $\Phi:X\to Y$ whose restriction to a submodule of X is an isomorphism, splits: indeed if $X_1 \subset X_2$ and $\Phi|X_1$ is an isomorphism then $\ker\Phi\cap X_1=0$ and $\ker\Phi+X_1=X$.

(4) Consider next the case of a surjection $f:M\to C$ from our module M to an indecomposable (hence local) summand C . According to (E) , C is isomorphic to some A_k , via some isomorphism Φ . Since C is local, $f(A_i)=C$ for some i , hence $\Phi f:A_i\to A_k$ is surjective, hence splits since A_k is A_i-projective for $i\neq k$ and d-continuous. As A_i is indecomposable, $\Phi f|A_i$ is an isomorphism hence so is $f|A_i$, and thus f splits by (3) .

(5) Finally consider an arbitrary surjection $f:M\to B$ onto a non-zero summand B of M . Due to (F) we may write $B=C\oplus D$, with projections π_C and π_D , where C is local and D has smaller direct-decomposition-length than B . Hence we may assume by induction that $\pi_D f:M\to D$ splits, ie. $M=M_1\oplus M_2$ such that $\pi_D f|M_1$ is isomorphic onto D and $M_2=\ker\pi_D f$.

Then $\pi_C f|M_2$ is again surjective: indeed for each $c\in C\subset B$ there is $m\in M$ with $f(m)=c$. Decompose $m=m_1+m_2\in M_1\oplus M_2$; then $0=\pi_D(c)=\pi_D f(m)$ $\pi_D f(m_1)$ since $m_2\in M_2=\ker\pi_D f$, hence $0=m_1$ since $\pi_D f|M_1$ is an isomorphism hence $m=m_2\in M_2$ and $c=\pi_C(c)\in\pi_C f(M_2)$.

Therefore by (4) , $\pi_C f|M_2$ splits, ie. $M_2=M_3\oplus M_4$ such that $\pi_C f|M_3$ is isomorphic onto C and $M_4 = \ker(\pi_C f|M_2)$. Then $f|M_1\oplus M_3$

is isomorphic to B : indeed if $f(m_1+m_3)=0$, then $\pi_D f(m_1)=0$ since $m_3 \in M_3 \subset M_2 = \ker \pi_D f$, hence $m_1=0$ since $\pi_D f|M_1$ is an isomorphism. But then $f(m_3)=0$, hence $\pi_C f(m_3)=0$ hence $m_3=0$ since $\pi_C f|M_3$ is an isomorphism, proving that the map $f|M_1 \oplus M_3$ is injective. Moreover if $m_4 \in M_4 = \ker(\pi_C f|M_2) \subset M_2 = \ker \pi_D f$, then $\pi_C f(m_4)=0$ and $\pi_D f(m_4)=0$ hence $f(m_4)=0$; therefore $B=f(M)=f(M_1 \oplus M_3 \oplus M_4)=f(M_1 \oplus M_3)$ and the map is also surjective.

Consequently since $M=(M_1 \oplus M_3) \oplus M_4$, f splits by the observation (3) , as required in condition (II) .

REFERENCES

1. G. Azumaya, M-projective and M-injective modules (preprint).
2. G. Azumaya, F. Mbuntum and K. Varadarajan, On M-projective and M-injective modules, Pacific J. Math. 59 (1975), 9-16.
3. Y. Miyashita, Quasi-projective modules, perfect modules, and a theorem for modular lattices, J. Fac. Sci. Hokkaido Univ. 19 (1966), 88-110.
4. S. Mohamed and T. Bouhy, Continuous modules (preprint; cf. Notices Amer. Math. Soc. 23 no. 5 (1076), A-478).
5. S. Mohamed and S. Singh, Generalizations of decomposition theorems known over perfect rings, J. Austral. Math. Soc. (to appear).
6. Y. Utumi, On continuous rings and self-injective rings, Trans. Amer. Math. Soc. 118 (1965), 158-173.
7. R. B. Warfield, A Krull-Schmidt theorem for infinite sums of modules, Proc. Amer. Math. Soc. 22 (1969), 460-465.

This research was supported in part by the NRC of Canada.

ON THE GABRIEL DIMENSION AND SUBIDEALIZER RINGS[1]

Friedhelm Hansen
Universität Bochum
D-4630 Bochum

and

Mark L. Teply[2]
University of Florida
Gainesville, Florida 32611

In this paper M will always denote a right ideal of a ring T with an identity element, and R will always denote a unital subring of T that contains M as a two-sided ideal. Such a subring R is called a subidealizer of M in T. To avoid trivialities we always assume that $M \neq T$.

We assume that the reader is familiar with the definition of the Gabriel dimension of a module that is given in [6], [7], [8], and [10]. For a module A, we use the notation G-dim A for the Gabriel dimension of A. Modules frequently carry subscripts to indicate which ring (R or T) and side that we are using; e.g. G-dim ${}_RA$ denotes the Gabriel dimension of the left R-module A,

1. This work was done independently by the two authors, who each learned of the other's work through communications with G. Krause.

2. The work of this author was supported by NSF grant MCS 77-01818.

and G-dim A_T denotes the Gabriel dimension of the right T-module A. We frequently use the following basic property [7, Lemma 1.3] of the Gabriel dimension without reference: if

$$0 \to A \to B \to C \to 0$$

is an exact sequence of modules, then

$$\text{G-dim } B = \max\{\text{G-dim } A, \text{ G-dim } C\},$$

provided that either side of the equation exists.

As usual, the symbol ω denotes the first infinite ordinal.

Techniques involving subidealizers have recently been employed to provide solutions to a number of ring theoretic questions; e.g. see [3], [4], [5], [11], [12], and [14]. In [9] and [10] conditions are obtained that guarantee the equality of the Gabriel dimensions of T and R; if B is a T-module, conditions are also obtained to insure the equality of the Gabriel dimension of B as a T-module and the Gabriel dimension of B as an R-module. In general, the corresponding dimensions do not coincide. Hansen [8] obtains bounds on the amount by which these dimensions may differ; in particular, the following results are obtained in [8].

THEOREM A. [8, Satz 3.1] If G-dim $_R(R/M) = \alpha$ and G-dim $_TB = \beta$, then

$$\text{G-dim } _RB \le \begin{cases} \alpha + (\beta - 1) & \text{if } \beta < \omega, \\ \alpha + \beta & \text{otherwise.} \end{cases}$$

Conversely, if G-dim ${}_T(T/TM) = \alpha$, G-dim ${}_R(R/M) = 1$, and G-dim ${}_RB = \gamma$ for a left T-module B, then

$$\text{G-dim }{}_TB = \begin{cases} \alpha + (\gamma-1) & \text{if } \gamma < \omega \text{ and } 1 \le \alpha, \\ \alpha + \gamma & \text{otherwise.} \end{cases}$$

THEOREM B. [8, Satz 4.1] If G-dim ${}_R(R/M) = \alpha$ and G-dim $B_T = \beta$, then

$$\text{G-dim } B_R \le \begin{cases} \alpha + (\beta-1) & \text{if } \beta < \omega, \\ \alpha + \beta & \text{otherwise.} \end{cases}$$

Conversely, if G-dim $(T/M)_T = \alpha$ and G-dim $B_R = \gamma$ for a right T-module B, then

$$\text{G-dim } B_T \le \begin{cases} \alpha + (\gamma-1) & \text{if } \gamma < \omega, \\ \alpha + \gamma & \text{otherwise.} \end{cases}$$

Except for the second part of Theorem A, these two results are very general. In section 1 we remove the restrictive hypothesis, G-dim ${}_R(R/M) = 1$, from the second part of Theorem A.

Hansen [8, page 591] also gives examples of commutative rings whose Gabriel dimensions are the upper bounds obtained in the first parts of Theorems A and B are the best possible. In section 2, we study some special situations and obtain some lower bounds for G-dim B_R when G-dim B_T is given. A result on valuation rings in section 2 shows that, for commutative rings, the bounds obtained in the second parts of Theorems A and B are also the best possible.

More generally, the inequalities in the second parts of Theorems A and B are indicative of what happens when M is not generative. (M is <u>generative</u> if TM = T.) However, for many uses of subidealizers of noncommutative rings, M is assumed to be generative and semimaximal (i.e. a finite intersection of maximal right ideals of T); e.g. see [3], [4], [5], [9], [11], and [12]. Whenever M is semimaximal and generative, results in the literature suggest that the dimensions of R and T should be closer together. Indeed, if M is generative, G-dim ${}_{T}T = \beta$, and G-dim ${}_{R}(R/M) = \alpha$, then Theorems A and [1, Theorem 1.1] predict that

$$\max\{\alpha,\beta\} \leq \text{G-dim } {}_{R}R = \text{G-dim } {}_{R}T \leq \begin{cases} \alpha + (\beta-1) & \text{if } \beta < \omega, \\ \alpha + \beta & \text{otherwise.} \end{cases}$$

A similar prediction is given by Theorem B and [1, Theorem 1.1] when G-dim $T_T = \beta$ and G-dim $(R/M)_R = \alpha$. In section 3 we give examples to show that, for $1 \leq \beta \leq \omega$ and any nonlimit ordinal α, G-dim ${}_{R}R$ and G-dim R_R can take on any given ordinal value between $\max\{\alpha,\beta\}$ and $\alpha + \beta - 1$; in these examples M is a generative, maximal right ideal of T. We also briefly explore the case $\beta = \omega$.

While Theorems A and B give bounds on the difference of the Gabriel dimensions of T and R, it would be nice to have an exact formula at least in interesting special cases. This has been done for a few special cases where equality of the dimensions occurs; see [9] and [10] as well as [8, Korollar 3.3 (iii) and Korollar 4.4 (v)]. An exact formula for a special case in which the maximum difference always occurs is provided by Lemma 3.2. An exact formula for the case where T is a valuation ring is given in Theorem 2.3. In Theorem 4.1 we give an exact formula suggested by commonly occuring matrix ring examples such as Example 3.1. In particular, we show that if M is a generative right ideal of T and M_T is a direct summand of T, then

$$(*) \quad \text{G-dim } R = \{\text{G-dim } T, \text{ G-dim } R/M\}$$

provided that either side of the equation exists. If the hypothesis that TM = T is dropped, then inequality can occur between the terms in (*) even though M_T is a direct summand of T.

1. The improvement of Theorem A.

Let β be a nonlimit ordinal. We recall from [7] that a module S is β-simple if (i) every proper homomorphic image of S has Gabriel dimension < β and (ii) S has no nonzero submodule of Gabriel dimension < β. A module is called Gabriel simple if it is β-simple for some β.

We are now ready for our key lemma.

LEMMA 1.1. Assume that $\text{G-dim}_T(T/TM) = \alpha$, that $X \in$ T-mod, and that Rx is a β-simple R-submodule of X for some nonlimit ordinal β. Then

$$\text{G-dim}_T(Tx) \leq \begin{cases} \alpha + (\beta-1), & \text{if } \beta < \omega \text{ and } 1 \leq \alpha, \\ \alpha + \beta & \text{otherwise.} \end{cases}$$

Proof. Let $\mathcal{T}$ be the smallest (not necessarily hereditary) torsion class of T-mod such that $T/TM \in \mathcal{T}$. Then ${}_TK \in \mathcal{T}$ if and only if, for each ${}_TJ \subsetneq K$, there exists a $k \in K - J$ such that $Mk \subseteq J$. Thus any $L \in \mathcal{T}$ satisfies G-dim ${}_TL \le \alpha$ by [7, Lemma 1.4]. Let $\mathcal{T}(Tx)$ denote the largest T-submodule of Tx in $\mathcal{T}$. In order to simplify our notation, we set $V = (Rx + \mathcal{T}(Tx))/\mathcal{T}(Tx)$ and $W = Tx/\mathcal{T}(Tx)$. We may assume that $V \ne 0$; for otherwise $Tx = \mathcal{T}(Tx)$, and hence

$$\text{G-dim}\ {}_T(Tx) \le \alpha \le \begin{cases} \alpha + (\beta-1), & \text{if } \beta < \omega \text{ and } 1 \le \alpha, \\ \alpha + \beta & \text{otherwise.} \end{cases}$$

Since $Mtx \subseteq Rx$ and $Mtx \not\subseteq \mathcal{T}(Tx)$ for each $tx \in Tx - \mathcal{T}(Tx)$, then V is an essential R-submodule of W.

We use induction on β = G-dim ${}_R(Rx)$.

Let $\beta = 1$. Since Rx is a simple R-module in this case and since ${}_RV$ is assumed to be an essential, nonzero R-submodule of W, then any nonzero T-submodule of W contains $x + \mathcal{T}(Tx)$. Thus Tx must be a simple T-module. Hence

$$\begin{aligned} \text{G-dim}\ {}_T(Tx) &\le \max\{\text{G-dim}\ {}_T(Tx),\ \ \text{G-dim}\ {}_TW\} \\ &\le \max\{\alpha, 1\} \\ &\le \begin{cases} \alpha + (\beta-1), & \text{if } \beta = 1 \text{ and } 1 \le \alpha, \\ \alpha + \beta & \text{otherwise.} \end{cases} \end{aligned}$$

Now let β be a nonlimit ordinal such that $\beta > 1$. Assume that the lemma is true for any γ-simple R-module, where γ is a nonlimit ordinal such that $\gamma < \beta$. We consider two subcases:

(i) $\mathcal{T}(Tx) \cap Rx = 0$, and (ii) $\mathcal{T}(Tx) \cap Rx \ne 0$.

Case (i). Here $Rx \cong V$, and hence ${}_R V$ is β-simple. First, we wish to show that

$$\text{G-dim}\ {}_T W \le \begin{cases} \alpha + (\beta-1), & \text{if } \beta < \omega \text{ and } 1 \le \alpha, \\ \alpha + \beta & \text{otherwise.} \end{cases}$$

Let $0 \ne {}_T Y \subseteq W$; then $V \cap Y \ne 0$. Moreover, if $x + T(Tx) \in Y$, then $Y = W$; whence G-dim ${}_T(W/Y) = 0$. Thus we may assume that $V \cap Y \ne V$. Since ${}_R V$ is β-simple, then $0 <$ G-dim ${}_R((V + Y)/Y) =$ G-dim ${}_R(V/(V \cap Y)) = \delta < \beta$. Let $(Rv + Y)/Y$ be a nonzero γ-simple R-submodule of $(V + Y)/Y$, where γ is a nonlimit ordinal such that $\gamma \le \delta < \beta$. By our induction hypothesis (as $(Rv + Y)/Y \subseteq W/Y)$, we obtain

$$\text{G-dim}\ {}_T((Tv + Y)/Y) \le \begin{cases} \alpha + (\gamma-1), & \text{if } \gamma < \omega \text{ and } \alpha \ge 1, \\ \alpha + \gamma & \text{otherwise} \end{cases}$$
$$< \begin{cases} \alpha + (\beta-1), & \text{if } \beta < \omega \text{ and } \alpha \ge 1, \\ \alpha + \beta & \text{otherwise.} \end{cases}$$

Hence

$$\text{G-dim}\ {}_T W \le \begin{cases} \alpha + (\beta-1), & \text{if } \beta < \omega \text{ and } \alpha \ge 1, \\ \alpha + \beta & \text{otherwise.} \end{cases}$$

In view of this inequality, we have

$$\begin{aligned} \text{G-dim}\ {}_T(Tx) &= \max\{\text{G-dim}\ {}_T(T(Tx)),\ \text{G-dim}\ {}_T W\} \\ &\le \max\{\alpha,\ \text{G-dim}\ {}_T W\} \\ &\le \begin{cases} \alpha + (\beta-1) & \text{if } \beta < \omega \text{ and } \alpha \ge 1, \\ \alpha + \beta & \text{otherwise.} \end{cases} \end{aligned}$$

Case (ii). Since Rx is β-simple and $\mathcal{T}(Tx) \cap Rx \neq 0$, then G-dim ${}_RV < \beta$. Let $0 \neq {}_TK \subseteq W$; so $0 \neq K \cap V \neq V$. Let C be any nonzero, γ-simple, cyclic R-submodule of $(V + K)/K$; we must have $\gamma < \beta$ and γ is a nonlimit ordinal. Since $C \subseteq W/K$, our induction hypothesis implies that

$$0 < \text{G-dim }_T(TC) \leq \begin{cases} \alpha + (\gamma-1), & \text{if } \gamma < \omega \text{ and } 1 \leq \alpha, \\ \alpha + \gamma & \text{otherwise;} \end{cases}$$
$$< \begin{cases} \alpha + (\beta-1), & \text{if } \beta < \omega \text{ and } 1 \leq \alpha, \\ \alpha + \beta & \text{otherwise.} \end{cases}$$

Hence every proper, nonzero T-homomorphic image W/K of W has a nonzero T-submodule TC such that

$$\text{G-dim }_T(TC) < \begin{cases} \alpha + (\beta-1), & \text{if } \beta < \omega \text{ and } 1 \leq \alpha, \\ \alpha + \beta & \text{otherwise.} \end{cases}$$

Hence

$$\text{G-dim }_TW \leq \begin{cases} \alpha + (\beta-1), & \text{if } \beta < \omega \text{ and } 1 \leq \alpha, \\ \alpha + \beta & \text{otherwise.} \end{cases}$$

We conclude that

$$\begin{aligned} \text{G-dim }_T(Tx) &= \max\{\text{G-dim }_T\mathcal{T}(Tx),\ \text{G-dim }_TW\} \\ &\leq \max\{\alpha,\ \text{G-dim }_TW\} \\ &\leq \begin{cases} \alpha + (\beta-1), & \text{if } \beta < \omega \text{ and } 1 \leq \alpha, \\ \alpha + \beta & \text{otherwise.} \end{cases} \end{aligned}$$

We are now ready for the main result of this section.

THEOREM 1.2. Let $B \in T$-mod. If G-dim ${}_T(T/TM) = \alpha$ and G-dim ${}_RB = \gamma$, then G-dim ${}_TB$ exists and

$$\text{G-dim }{}_TB \le \begin{cases} \alpha + (\gamma-1), & \text{if } \gamma < \omega \text{ and } 1 \le \alpha, \\ \alpha + \gamma & \text{otherwise.} \end{cases}$$

Proof. By [7, Lemma 1.4], it is sufficient to show that, for each ${}_TC \subseteq B$, there exists ${}_TD$ such that $C \subsetneq D \subseteq B$ and

$$\text{G-dim }{}_T(D/C) \le \begin{cases} \alpha + (\gamma-1), & \text{if } \gamma < \omega \text{ and } 1 \le \alpha, \\ \alpha + \gamma & \text{otherwise.} \end{cases}$$

Since G-dim ${}_R(B/C) \le \alpha$, then there exists a nonzero, β-simple, cyclic R-submodule $(Rx + C)/C$ of B/C, where β is a nonlimit ordinal such that $1 \le \beta \le \alpha$. By Lemma 1.1,

$$\text{G-dim }{}_T((Tx + C)/C) \le \begin{cases} \alpha + (\gamma-1), & \text{if } \gamma < \omega \text{ and } 1 \le \alpha, \\ \alpha + \gamma & \text{otherwise.} \end{cases}$$

Hence we can choose $D = Tx + C$.

COROLLARY 1.3. If G-dim ${}_T(T/TM) = \alpha$ and G-dim ${}_RR = \gamma$, then G-dim ${}_TT$ exists and

$$\text{G-dim }{}_TT \le \begin{cases} \alpha + (\gamma-1), & \text{if } \gamma < \omega \text{ and } 1 \le \alpha, \\ \alpha + \gamma & \text{otherwise.} \end{cases}$$

COROLLARY 1.4. Let $B \in T$-mod. If M is generative and G-dim ${}_RB$ exists, then G-dim ${}_TB$ exists and

$$\text{G-dim }{}_TB \le \text{G-dim }{}_RB.$$

If $B \in$ T-mod and G-dim ${}_R B$ exists, it is sometimes possible to show that G-dim ${}_T B \leq$ G-dim ${}_R B$ without assuming that G-dim ${}_T(T/TM)$ exists. Suppose that no T-subfactor of B is annihilated by M. If Rx is any β-simple submodule of a nonzero T-subfactor X of ${}_T B$, then $T(Tx) = 0$ (in the notation of the proof of Lemma 1.1). Thus the proof of Lemma 1.1 can be modified and simplified to show that G-dim ${}_T(Tx) \leq$ G-dim ${}_R(Rx)$. But then the proof of Theorem 1.2 yields the following result.

THEOREM 1.6. Let $B \in$ T-mod such that no T-subfactor of B is annihilated by M. If G-dim ${}_R B$ exists, then G-dim ${}_T B$ exists and

$$\text{G-dim } {}_T B \leq \text{G-dim } {}_R B.$$

2. Lower bounds and valuation rings

In this section, we study the transfer of the Gabriel dimension in some special situations.

If $B \in$ mod-T and G-dim B_T is given, then our first result, Theorem 2.1, gives a lower bound for G-dim B_R. The case $\gamma = 0$, which is always true, was proved in [8, Satz 4.2].

THEOREM 2.1. Let G-dim $(T/M)_T = 1$, and let γ be an ordinal number such that G-dim $B_R \geq \gamma$ for each nonzero $B \in$ mod T. If G-dim $B_T = \beta$ and B_R has Gabriel dimension, then

$$\text{G-dim } B_R \geq \begin{cases} \gamma + (\beta-1) & \text{if } \beta < \omega, \\ \gamma + \beta & \text{otherwise.} \end{cases}$$

Proof. The case $\beta = 1$ is obvious; so we use induction on β.

$1 < \beta < \omega$. Observe that B_T may be assumed to be β-simple. Then the hypotheses force $bM \neq 0$ for all nonzero $b \in B$. Since B_R has Gabriel dimension, there exists a δ-simple (nonzero) R-submodule A of B. Then AM is both δ-simple as an R-module and β-simple as a T-module. By induction there is a proper T-factor AM/C of AM with G-dim $(AM/C)_R \geq \gamma + (\beta-2)$. Hence G-dim $B_R \geq$ G-dim $(AM)_R \geq \gamma + (\beta-1)$ as desired.

$\beta = \omega$. G-dim $B_R = \sup\{$G-dim $A_R | A_T$ is a δ-simple T-subfactor of B_T, $\delta < \omega\} \geq \sup\{\gamma + (\delta-1) | \delta < \omega\} = \gamma + \omega$.

$\beta \geq \omega + 1$. In this case G-dim $B_R = \sup\{$G-dim $A_R | A_T$ is a δ-simple T-subfactor of B_T, $\omega + 1 \leq \delta \leq \beta\}$. It follows that we need only prove the result for the case where B_T is δ-simple, $\omega + 1 \leq \delta \leq \beta$. The argument in the case $1 < \beta < \omega$ shows that we may also assume B_R to be Gabriel simple. By induction, we have $\sup\{$G-dim $F_R | F$ is a proper T-factor module of $B_T\} \geq \sup\{\gamma + \mu | \omega \leq \mu \leq \delta-1\} = \gamma + \delta - 1$. Since B_R is Gabriel simple, we obtain G-dim $B_R \geq \gamma + \delta$ as desired.

We note that Example A in [8] gives rings that satisfy the conditions of Theorem 2.1 for any given pair β and γ of nonlimit ordinals.

The rings constructed in Example A of [8] are valuation rings. We study the valuation ring situation further to show that the bounds obtained in the second parts of Theorems A and B are the best possible.

We need the following preliminary result.

LEMMA 2.2. Let T be a valuation ring, let G-dim $R/M = \alpha'$, and let G-dim $(T/M)_T = \alpha$. If $B \in$ mod-T is Gabriel simple as a T-module and G-dim $B_R = \delta$ for some $\delta \leq \alpha'$, then G-dim $B_T \leq \alpha$.

Proof. The module B has a ρ-simple R-submodule A for some $\rho \leq \delta \leq \alpha'$. A nonzero cyclic submodule R/K of A is also ρ-simple. If KT were properly contained in M, then the ρ-simplicity of R/K implies that

$$\alpha' = \text{G-dim } R/M < \text{G-dim } R/K = \rho \leq \alpha',$$

which is a contradiction. Since T is a valuation ring, we must have $M \subseteq KT$. Since B_T is Gabriel simple, we now have

$$\text{G-dim } B_T \leq \text{G-dim } (T/KT)_T \leq \text{G-dim } (T/M)_T = \alpha$$

as desired.

We are now ready for our main result on valuation rings.

THEOREM 2.3. Let α' and τ be nonlimit ordinals. Let T be a valuation ring with G-dim $T = \tau$, and let G-dim $R/M = \alpha'$. Suppose that G-dim $(T/M)_T = \alpha < \tau$. Then the following statements hold.

(1) If $B \in$ mod-T has G-dim $B_R = \gamma > \alpha'$, then G-dim $B_T = \alpha + \xi$, where ξ is defined by $\alpha' + \xi = \gamma$.

(2) If $B \in$ mod-T has G-dim $B_T = \beta > \alpha$, then G-dim $B_R = \alpha' + \mu$, where μ is defined by $\alpha + \mu = \beta$.

Proof. (1). We use induction on ξ.

$\xi = 1$. Suppose that G-dim $B_T \leq \alpha$, and choose an α''-simple cyclic T-subfactor T/L of B_T, where $\alpha'' \leq \alpha$. Since T/L is

α''-simple and $\alpha'' \leq \alpha$, we cannot have L properly contained in M. Since T is a valuation ring, we must have $M \subseteq L$. Thus T/L is a canonical R/M-module; whence G-dim $(T/L)_R \leq \alpha'$. Since B_R is contained in the torsion class generated by the set of all its α''-simple, cyclic T-subfactors, it follows that G-dim $B_R \leq \alpha' < \gamma$, which contradicts our hypothesis. Therefore G-dim $B_T \geq \alpha + 1$.

We finish the case $\xi = 1$ by showing that G-dim $B_T \geq \alpha + 2$ is false when $\xi = 1$. In view of Lemma 2.1, we may assume that B_T is γ'-simple, where $\gamma' \geq \alpha + 2$. Now B_T contains a δ-simple R-submodule A, and the nonzero module AM is γ'-simple over T and δ-simple over R, where $\delta \leq \alpha' + 1$. In particular, for each proper T-factor module G of AM, we have G-dim $G_R \leq \alpha'$. It follows from Lemma 2.1 that every Gabriel simple T-subfactor of G has Gabriel dimension $\leq \alpha$; hence G-dim $G_T \leq \alpha$ also. Since AM is Gabriel simple, this implies that

$$\alpha + 2 \leq \gamma' = \text{G-dim } (AM)_T \leq \alpha + 1,$$

which is a contradiction.

$\xi > 1$. Let A be a γ'-simple R-subfactor of B, where $\alpha' + 2 \leq \gamma' \leq \gamma$. Again A contains a nonzero submodule E_T that is γ'-simple as an R-module and Gabriel simple as a T-module. Now

$$\sup\{\text{G-dim } F_R \mid F_R \text{ is a proper factor module of } E\} = \gamma' - 1,$$

Since $FM \neq 0$, induction yields

$$\sup\{\text{G-dim } F_T | F_T \text{ is a proper factor module of } E\} = \alpha + (\rho-1),$$

where ρ is defined by $\alpha' + \rho = \gamma'$. Since E_T is Gabriel simple, we have G-dim $E_T = \alpha + \rho$. Now the set of all subfactors of the type E_T generate a hereditary torsion class containing B, and $\sup\{\gamma' | \gamma' = \text{G-dim } E_R \text{ for some } E_T\} = \gamma$. Hence G-dim $B_T = \sup\{\alpha + \rho | \alpha' + \rho = \gamma', \gamma' \text{ is a nonlimit ordinal}, \gamma' \leq \gamma\} = \alpha + \xi$.

(2) follows from (1) since the values are uniquely determined.

Remarks. (1) By setting $\alpha' = 1$ in Theorem 2.3, we see that the rings constructed in Example A of [8] show that the bounds attained in the second parts of Theorems A and B are the best possible.

(2) Roughly speaking, Theorem 2.3 means that G-dim R is close to G-dim R/M if and only if G-dim T is close to G-dim T/M (where T is a valuation ring). Theorem 2.3 also gives complete information about the classical Krull dimension of subrings R of valuation rings T, where R contains a nonzero ideal M of T. (R need not be a valuation ring.)

(3) In the special case of valuation rings T, the exact Gabriel dimension of R depends only on the parameters

$$\alpha = \text{G-dim } (T/M)_T, \ \tau = \text{G-dim } T_T, \text{ and } \alpha' = \text{G-dim } R/M,$$

while the exact Gabriel dimension of T depends only on

$$\alpha,\ \alpha',\ \text{and}\ \rho = \text{G-dim}\ R_R.$$

This is of course not true in general cases.

For example, let T' be a simple ring with G-dim T' = 3 (e.g. a Weyl algebra), let M' be a right ideal of T' with G-dim T'/M' = 2, and let R' be a subring of T' such that $R' \supseteq M'$ and G-dim R'/M' = 1 (say R' = center T' + M'). Then by [8, Satz 4.1], G-dim R' $\leq$ 1 + (3-1) = 3; in fact, we must have equality because of a localization-like situation for $R' \subseteq T'$ on the right. If we choose the same parameters for a valuation ring T, we obtain G-dim R = 2.

On the other hand, let

$$\alpha = 1,\ \alpha' = 2,\ \text{and}\ \rho = 3.$$

Then G-dim T = 2, when T is a valuation ring. Let $T' = T'' \oplus Q$, where T" is any ring with G-dim T" = 3 and Q is the rational numbers. If $R' = T'' \oplus Z$ (Z = integers) and $M' = T'' \oplus 0$, then we obtain the same parameters

$$\alpha = 1,\ \alpha' = 2,\ \rho = 3,$$

although G-dim T' = 3.

3. Examples for Theorems A and B when M is generative.

Most of the examples that are given in [8] to illustrate that the bounds in the first parts of Theorems A and B are the best possible are commutative. However, in the literature (e.g. see [3], [4], [5], [9], [11], [12], and [14]) most of the right ideals M that are considered in forming subidealizers are generative (i.e. TM = T); thus the ring must be noncommutative. Results in the literature strongly suggest that if M is semimaximal (i.e. a finite intersection of maximal right ideals) and generative, then the properties of T and R are closely related. Thus we might expect to improve the bounds of Theorems A and B when M is maximal and generative. The purpose of this section is to show that this is not the case when G-dim $T < \omega$. In particular, we show in Example 3.4 that, given any ordinal k with $1 \le k < \omega$ and any nonlimit ordinal α, there exists a subidealizer R of a generative, maximal right ideal M satisfying the following conditions:

(i) G-dim ${}_T T = k =$ G-dim T_T,

(ii) G-dim ${}_R(R/M) = \alpha =$ G-dim $(R/M)_R$, and

(iii) G-dim ${}_R R = \gamma =$ G-dim R_R, where γ is any specified ordinal such that

$$\max\{k,\alpha\} \le \gamma \le \alpha + k - 1.$$

After doing this construction, we briefly examine the case where G-dim ${}_T T = \omega$.

The case where G-dim ${}_T T = 1 =$ G-dim T_T is easily handled.

EXAMPLE 3.1. Let α be a nonlimit ordinal. Let D be a commutative integral domain such that G-dim ${}_D D = \alpha$. (See [6, Theorem 9.8] and [7, Corollary 3.5] for the existence of such a domain D.) Let F be the quotient field of D. Now let

$$T = \begin{pmatrix} F & F \\ F & F \end{pmatrix}, \quad M = \begin{pmatrix} 0 & 0 \\ F & F \end{pmatrix}, \quad \text{and } R = \begin{pmatrix} D & 0 \\ F & F \end{pmatrix}.$$

Then M is a generative, maximal right ideal of T, G-dim ${}_TT = 1 =$ G-dim T_T, G-dim ${}_R(R/M) = \alpha =$ G-dim $(R/M)_R$, and G-dim ${}_RR = \alpha =$ G-dim R_R. (See section 4 for the methods of the computations of the Gabriel dimensions of R.)

We now turn to the case where $1 <$ G-dim ${}_TT =$ G-dim T_T. We will need the following result.

LEMMA 3.2. Let T be a hereditary noetherian domain. If G-dim ${}_R(R/M) = \alpha$, then G-dim ${}_RR = \alpha + 1$. If G-dim $(R/M)_R = \alpha$, then G-dim $R_R = \alpha + 1$.

Proof. By [8, Korollar 3.2 (i)], G-dim ${}_RR \leq$ G-dim ${}_R(R/M) + 2-1 =$ G-dim ${}_R(R/M) + 1$. Since T is an integral domain, then so is R; hence R is Gabriel simple by [7, Proposition 3.3]. Thus G-dim ${}_RR \geq$ G-dim ${}_R(R/M) + 1$.

A similar proof works for the right Gabriel dimension.

If T is a hereditary noetherian domain, then G-dim ${}_TT = 2 =$ G-dim T_T. If M is an essential, generative, maximal right ideal of T, then the largest subidealizer R (i.e. the <u>idealizer</u> of M in T) of M in T is also a hereditary noetherian domain, and R/M is a semisimple artinian ring. Hence G-dim ${}_RR = 2 =$ G-dim R_R and G-dim ${}_R(R/M) = 1 =$ G-dim $(R/M)_R$. See [4] for details.

For each nonlimit ordinal $\alpha \geq 2$, [10, Examples1.8 and 3.5] give a hereditary noetherian domain T of characteristic 0 with a subidealizer R of an essential, generative, maximal right ideal M such that G-dim ${}_R(R/M) = \alpha =$ G-dim $(R/M)_R$.

Thus the maximum values predicted in the first parts of Theorem A and B can be obtained whenever α is not a limit ordinal and G-dim ${}_TT \leq 2$. Our next example shows that the maximum value can also be obtained when $3 \leq$ G-dim ${}_TT =$ G-dim $T_T < \omega$.

EXAMPLE 3.3. Let α be a nonlimit ordinal. Let D be a hereditary noetherian domain of characteristic 0, and let K be an essential, generative, maximal right ideal of D. Let P be a subidealizer of K in D such that G-dim ${}_P(P/K) = \alpha =$ G-dim $(P/K)_P$. Let $T = D[x_1, x_2, \ldots, x_n]$, where the indeterminates commute. Clearly, $M = K + \sum_{i=1}^{n} Tx_i$ is an essential, generative, maximal right ideal of T, and $R = P + \sum_{i=1}^{n} Tx_i$ is a subidealizer of M in T. Moreover,

(a) G-dim ${}_TT = n + 2 =$ G-dim T_T,

(b) G-dim ${}_R(R/M) = \alpha =$ G-dim $(R/M)_R$,

(c) G-dim ${}_RR = \alpha +$ G-dim ${}_TT - 1$, and

(d) G-dim $R_R = \alpha +$ G-dim $T_T - 1$.

Proof. We prove the results for the left Gabriel dimension; the proof for the right Gabriel dimension is similar.

First, note that G-dim ${}_R(R/M) =$ G-dim ${}_P(P/K) = \alpha$.

Let K-dim A denote the Krull dimension of a module A in the sense of [6]. Since T is noetherian, then K-dim ${}_TT =$ K-dim ${}_DD + n = 1 + n$ by a result of Rentschler and Gabriel (see [6, Theorem 9.2]). By [7, Proposition 2.3] G-dim ${}_TT = n + 2$.

By Theorem A, R has Gabriel dimension. Since R is an integral domain, R is a Gabriel simple module by [7, Proposition 3.3]. Hence G-dim ${}_R R \geq$ G-dim ${}_R(R/Tx_n) + 1 =$ G-dim ${}_{P+\sum_{i=1}^{n-1} Tx_i}(P + \sum_{i=1}^{n-1} Tx_i) + 1$. Continuing inductively, we obtain the inequality G-dim ${}_R R \geq$ G-dim ${}_P P + n$. By Lemma 3.2 we now have G-dim ${}_R R \geq (\alpha+1) + n = \alpha +$ G-dim ${}_T T - 1$; so in view of Theorem A, we obtain G-dim ${}_R R = \alpha +$ G-dim ${}_T T - 1$.

We now come to the main result of this section.

EXAMPLE 3.4. Let $0 \leq n \leq m < \omega$, and let α be a nonlimit ordinal. Then there exists a subidealizer R of an essential, generative, maximal right ideal M of a ring T of characteristic 0 such that

(i) G-dim ${}_T T = m + 1 =$ G-dim T_T,

(ii) G-dim ${}_R(R/M) = \alpha =$ G-dim $(R/M)_R$, and

(iii) G-dim ${}_R R = \max\{\alpha + n, m + 1\} =$ G-dim R_R.

Proof. Let H be an essential, generative, maximal right ideal of a ring F of characteristic 0, and let I be a subidealizer of H in F such that (a) G-dim ${}_F F = n + 1 =$ G-dim F_F, (b) G-dim ${}_I(I/H) = \alpha =$ G-dim $(I/H)_I$, and (c) G-dim ${}_I I = \alpha + n =$ G-dim I_I. We can do this by the following methods: (1) if $n = 0$, we use Example 3.1; (2) if $n = 1$, we use the discussion preceeding Example 3.3; (3) if $n > 1$, we use Example 3.3. Let J be any ring such that G-dim ${}_J J = m + 1 =$ G-dim J_J. Let

$T = F \dot{+} J$ (ring direct sum), and let $M = H \oplus J$. Then M is also an essential, generative, maximal right ideal of T, and G-dim $_T T = \max\{\text{G-dim } _F F, \text{G-dim } _J J\} = m + 1 = \max\{\text{G-dim } F_F, \text{G-dim } J_J\} = \text{G-dim } T_T$. Now set $R = I \dot{+} J$. Then R is a subidealizer of M in T, and G-dim $_R(R/M) = \text{G-dim } _I(I/H) = \alpha = \text{G-dim } (I/H)_I = \text{G-dim } (R/M)_R$. Moreover, G-dim $_R R = \max\{\text{G-dim } _I I, \text{G-dim } _J J\} = \max\{\alpha + n, m + 1\} = \max\{\text{G-dim } I_I, \text{G-dim } J_J\} = \text{G-dim } R_R$.

We now proceed to the case where G-dim $_T T = \omega$.

EXAMPLE 3.5. Let $0 \leq m < \omega$, and let α be a nonlimit ordinal. Then there exists a subidealizer R of an essential, generative, maximal right ideal M of a ring T such that

(i) G-dim $_T T = \omega = \text{G-dim } T_T$

(ii) G-dim $_R(R/M) = \alpha = \text{G-dim } (R/M)_R$, and

(iii) G-dim $_R R = \max\{\alpha + m, \omega\} = \text{G-dim } R_R$.

Proof. For each ordinal $n < \omega$, we use Example 3.4 to obtain a subidealizer R_n of an essential, generative, maximal right ideal of M_n of a ring T_n of characteristic 0 such that (a) G-dim $_{T_n}(T_n) = n + 1 = \text{G-dim } (T_n)_{T_n}$, (b) G-dim $_{R_n}(R_n/M_n) = \alpha = \text{G-dim } (R_n/M_n)_{R_n}$, and (c) G-dim $_{R_n}(R_n) = \alpha + n = \text{G-dim } (R_n)_{R_n}$. Let T be the subring of $\prod_{n<\omega} T_n$ generated by the identity element 1_T and $\bigoplus_{n<\omega} T_n$. Since $T/(\bigoplus_{n<\omega} T_n)$ is isomorphic to the ring of integers, then we must have G-dim $_T T = \omega = \text{G-dim } T_T$. (Note that this is the first use we have made of the fact that the characteristic of T is 0.) Let e_m be the element of $\prod_{n<\omega} T_n$

having 1 as its m-th coordinate and 0 for all of its other coordinates. Then $1_T - e_m \in T$. Now let

$$M = M_m \oplus T(1_T - e_m) \text{ and } R = R_m \oplus T(1_T - e_m).$$

Then M is an essential, generative, maximal right ideal of T, R is a subidealizer of M in T, and G-dim $_R(R/M)$ = G-dim $_{R_m}(R_m/M_m)$ = α = G-dim $(R_m/M_m)_{R_m}$ = G-dim $(R/M)_R$. Since $T(1_T - e_m)/(\bigoplus_{\substack{n<\omega \\ n\neq m}} T_n)$ is isomorphic to the ring of integers, it follows that G-dim $_RR = \max\{\alpha + m, \omega\}$ = G-dim R_R.

EXAMPLE 3.6. Given any nonlimit ordinal α, there exists a subidealizer R of an essential right ideal M of a ring T such that

(i) G-dim $_TT = \omega$ = G-dim T_T,

(ii) G-dim $_R(R/M) = \alpha$ = G-dim $(R/M)_R$, and

(iii) G-dim $_RR = \alpha + \omega$ = G-dim R_R.

Proof. Let T_n, T, M_n, and R_n be defined in Example 3.5, but choose $M = \bigoplus_{n<\omega} M_n$. Then M is an essential right ideal of T, but M is neither semimaximal nor generative. Let R be the subring of T generated by the identity element 1_T of T and $\bigoplus_{n<\omega} R_n$. Then R is a subidealizer of M in T. Since $R/(\bigoplus_{n<\omega} R_n)$ is isomorphic to the ring of integers, it follows that G-dim $_R(R/M) = \alpha$ and G-dim $_RR = \alpha + \omega$.

In Example 3.4 T is an integral domain, while T is not a domain in Example 3.5. Thus we raise the following question: what ordinal values can G-dim $_RR$ be if T is an integral domain?

4. The case when M is a direct summand.

Our examples in previous sections show that in general G-dim ${}_R R$ and dim R_R may take on any of the values predicted by Theorems A and B. It would be nice to obtain an exact formula for these dimensions. We can do this in some relatively elementary cases already; e.g. see Theorem 2.3 and Lemma 3.2 as well as results from [9] and [10]. The result of this section investigates another special case, namely the case where M_T is a direct summand of T.

THEOREM 4.1. Let M = eT for some idempotent $e \in R$. Then

$$\text{G-dim } R \leq \max\{\text{G-dim } T, \text{ G-dim } R/M\}$$

on the left and on the right. Furthermore, equality holds if M is generative.

Proof. Since M is an ideal of R, it follows that $(1 - e)Re = 0$. By [2, Theorem 2.3] R has the form

$$R \cong \begin{pmatrix} eRe & eR(1-e) \\ 0 & (1-e)R(1-e) \end{pmatrix}$$

It is easy to see that there exists an isomorphism between the rings R/M and (1-e)R(1-e). Since

$$\begin{pmatrix} 0 & eR(1-e) \\ 0 & 0 \end{pmatrix}^2 = 0,$$

then for $S = eRe \oplus (1-e)R(1-e)$ we have G-dim R = G-dim S = G-dim $(eTe \oplus R/M)$ = max{G-dim eTe, G-dim R/M} on the left and on the right. It is known that mod-eTe (respectively, eTe-mod) is equivalent to a quotient category of mod-T (respectively, T-mod); hence G-dim eTe $\leq$ G-dim T on the left and on the right. Thus we have

$$(\Delta) \quad \text{G-dim } R \leq \max\{\text{G-dim } T, \text{G-dim } R/M\}$$

on the left and on the right as desired. Moreover, if T = TM = TeT, then T and eTe are Morita equivalent; so G-dim eTe = G-dim T, and we obtain equality in (Δ).

REFERENCES

1. E. P. Armendariz and J. W. Fisher, Idealizers in rings, J. Algebra 39(1976), 551-562.

2. S. U. Chase, A generalization of the ring of triangular matrices, Nagoya Math. J. 18(1961), 13-25.

3. K. R. Goodearl, Idealizers and nonsingular rings, Pacific J. Math. 48(1973), 395-402.

4. ______________, Localization and splitting in hereditary noetherian prime rings, Pacific J. Math. 53(1974), 137-151.

5. ______________, Subrings of idealizer rings, J. Algebra 33(1975), 405-429.

6. R. Gordon and J. C. Robson, Krull dimension, Mem. Amer. Math. Soc. 133(1973).

7. ___________________________, The Gabriel dimension of a module, J. Algebra 29(1974), 459-473.

8. F. Hansen, Schranken für die Gabriel- und Krull-dimension Idealisator-ähnlicher Ringerweiterungen, Arch. Math. 28(1977), 584-593.

9. G. Krause, Krull dimension and Gabriel dimension of idealizers of semimaximal left ideals, J. London Math. Soc. (2) 12(1976), 137-140.

10. G. Krause and M. L. Teply, The transfer of the Krull dimension and the Gabriel dimension to subidealizers, Canad. J. Math. 29(1977), 874-888.

11. J. C. Robson, Idealizers and hereditary noetherian prime rings, J. Algebra 22(1972), 45-81.

12. ____________, Simple noetherian rings need not have unity elements, Bull. London Math. Soc. 7(1975), 269-270.

13. B. Stenström, Rings and modules of quotients, Lecture Notes in Math 237, Springer-Verlag, Berlin, 1971.

14. M. L. Teply, Prime singular-splitting rings with finiteness conditions, Noncommutative ring theory - Kent State 1975, Lecture Notes in Math. 545, Springer-Verlag, Berlin, 1976, 173-194.

15. ___________, On the transfer of properties to subidealizer rings, Communications in Algebra 5(1977), 743-758.

Big and Small Cohen-Macaulay Modules[1]

by Melvin Hochster[2]
University of Michigan
Ann Arbor, Michigan 48109

1. Introduction. Throughout this paper, rings are commutative, with identity, and modules are unital. The first part of the paper (§§1-4) is a brief survey of the present state of knowledge concerning existence of big and small C-M (≡ Cohen-Macaulay) modules, with an even briefer survey of their relationship to some other open questions in the homological theory of commutative rings. The presentation is intended for algebraists who are not specialists in commutative Noetherian rings.

The second part (§§5-6) contains some new results related to these questions, and it is assumed that the reader has much more background in commutative rings.

2. Big and small Cohen-Macaulay modules. Let (R,m) be a local ring (≡ Noetherian ring with <u>unique</u> maximal ideal m).

---

[1]The first part of this paper is based on a talk given by the author at the Special Session on Module Theory at the 747th meeting of the A.M.S., in Seattle, Washington, on August 17, 1977.

[2]The author was supported in part by a grant from the National Science Foundation.

Recall that $x_1, \ldots, x_n \in m$ is called a <u>system of parameters</u> (s.o.p. for short) if it is a sequence of minimal length such that for some t, $m^t \subset (x_1,\ldots,x_n)R$. In this case, n = Krull dim R. In fact, $x_1, \ldots, x_r$ is part of a s.o.p. $\Leftrightarrow$ $\dim R/(x_1,\ldots,x_r)R = \dim R-r$. Also recall that $x_1, \ldots, x_n$ in a ring R is called a <u>regular</u> sequence on an R-module E or an E-<u>sequence</u> if

1) $(x_1,\ldots,x_n)E \neq E$
2) For each i, $1 \leq i \leq n$, x_i is not a zerodivisor on $E/(x_1,\ldots,x_{i-1})E$.

We shall say that E is a <u>big</u> <u>C-M</u> <u>module</u> for the local ring (R,m) if R has a s.o.p. $x_1, \ldots, x_n$ which is an E-sequence. "Big" refers to the fact that E <u>need not be finitely generated</u>. If E <u>is</u> finitely generated then it will automatically be true that every s.o.p. is an E-sequence. We call such an E a <u>small</u> C-M module.

The main conjectures are:

(2.1) Conjecture. <u>If</u> R <u>is a local ring then</u> R <u>has a big</u> C-M <u>module</u>.

(2.2) Conjecture. <u>If</u> (R,m) <u>is a complete local ring then</u> R <u>has a small</u> C-M <u>module</u>.

[(R,m) is "complete" means that R is complete in the m-adic metric.]

(2.3) Remarks. a) Both (2.1) and (2.2) can be reduced

easily to the case of complete local normal (i.e., integrally closed) domains.

b) (2.2) is false if the completeness assumption on R is dropped (there are examples in [H_1] based on [FR], [N]). There is more discussion of the non-complete case later.

c) If R is a complete domain, R can be represented as a module-finite algebra extension of A , where A is a formal power series ring over a field or complete discrete rank one valuation ring. When R is represented in this way, a finitely generated R-module $M \neq 0$ is a small C-M module if and only if M is A-free. It follows that in this case R has a small C-M module if and only if it can be embedded in $M_n(A)$, the ring of n by n matrices over A , for some n , so as to extend the embedding of A as the scalar matrices.

In the same situation P. Griffith has shown that R has a big C-M module if and only if R has a nonzero countably generated R-module which is A-free [Gr].

3. The known results on Cohen-Macaulay modules.

(3.1) Theorem. *If R is a local ring which contains a field, then R has a big Cohen-Macaulay module.*

This is proved in [H_5] (see also [E], [H_3], [H_4], [H_6], [H_7]). The idea of the proof is interesting: roughly speaking, one shows that certain equations obstruct the existence of big Cohen-Macaulay modules. Using Artin approximation (see [Ar_1], [Ar_2], and §5 below) one reduces to the case of local rings of algebras finitely generated over a field, then over $\mathbb{Z}$, and, finally, over a finite field. In char. $p > 0$ one can apply the Frobenius homomorphism repeatedly to the equations to obtain a contradication.

(3.2) Theorem. _If_ R _is complete, local, and_ $\dim R \le 2$, _then_ R _has a small Cohen-Macaulay module. If_ R _is local and_ $\dim R \le 2$ _then_ R _has a big_ C-M _module._

Cf. [H_1], [LV]. Several proofs are known.

The question of existence of small C-M modules is almost completely open if $\dim R \ge 3$, while the question of existence of big C-M modules is likewise open if $\dim R \ge 3$ and R does not contain a field. The following result on small C-M modules is due to Hartshorne-Peskine-Szpiro (see the unpublished preprint [PS_1]): the proof is given in [H_6].

(3.3) Theorem. _Let_ R _be a domain which is a finitely generated graded_ K-_algebra over a field_ K _of characteristic_ $p > 0$ _and suppose_ R _has a graded torsion-free finitely generated module_ M _which is Cohen-Macaulay when localized at any prime except possibly the homogeneous maximal ideal of_ R .

Then R _has a small Cohen-Macaulay module_ E .

In particular, the result applies to graded domains R _if_

a) $\dim R \leq 3$.

b) R *is generated by its 1-forms over* K, *and* Proj (R), *the projective variety associated with* R , *is Cohen-Macaulay*.

Remarks. In case a), let M be the integral closure of R . In case b), let $M = R$. See $[H_6]$.

One approach to the mixed characteristic case (R does not contain a field) is to make use of Witt vectors. We shall not give details here, but for the purpose of showing the existence of big C-M modules in some such cases it helps to know the following result $[H_6]$:

(3.4) Theorem. *Let* R *be a local ring of char.* $p > 0$. *Then there exists a commutative* R-*algebra without unit* E *such that*

1) E *is a big* C-M *module for* R , *and*

2) *the Frobenius homomorphism* $F : E \to E$ *is an automorphism, i.e.,* E *has no nonzero nilpotents and every element has a* p^{th} *root*.

The point is that the (generalized) Witt vectors over E form a big C-M module over rings embeddable, in a "good" way, in the (generalized) Witt vectors over an enlargement of R . See §6 for a result on the nonexistence of certain C-M algebras.

4. Homological conjectues.

We shall list quickly several homological conjectures. All of these are implied by the existence of big Cohen-Macaulay modules in the general case, and all may be proved in the case where R contains a field using the theorem that big C-M modules exist in that case. In several instances we have, for simplicity, stated a more "special" form of the conjecture involved, but which is known to be equivalent to the more general form.

For background on the homological theory of local rings, we refer the reader to [AB_1], [AB_2], [AB_3], [B], [GH], [H_5], [K], [M], [N], [Re], [S_1], and [S_2], while for more detailed information about the specific homological questions raised here, we refer the reader to [Au_1], [Au_2], [EE], [FFGR], [F_1], [F_2], [F_3], [H_1] - [H_8], inclusive, [Iv], [PS_1], [PS_2], [PS_3], and [Ro].

(4.1) Conjecture (M. Auslander's zerodivisor conjecture). Let R be a local ring and $M \neq 0$ a finitely generated module of finite projective dimension. Suppose $x \in R$ is a zerodivisor on R. Then x is a zerodivisor on M.

A ring is called Cohen-Macaulay if each of its local rings is a C-M module over itself.

(4.2.) Conjecture (Bass' question). Let R be a local ring and $T \neq 0$ a finitely generated R-module. Suppose T has *finite injective dimension. Then* R *is Cohen-Macaulay.* [See [B], [PS_2].]

(4.3) Conjecture (Peskine-Szpiro intersection conjecture). *Let* (R,m) *be a local ring,* M, N *nonzero finitely generated* R-*modules, and suppose* $M \otimes N$ *is killed by a power of* m . *Then*

$$\text{Krull dim}(R/\text{Ann}_R N) \leq \text{pd}_R M ,$$

where "pd" *denotes projective dimension.* [See [PS_2].]

(4.4) Conjecture (homological height conjecture). *Let* $R \to S$ *be a homomorphism of Noetherian rings, let* $M \neq 0$ *be a finitely generated* R-*module, let* $I = \text{Ann}_R M$, *and let* Q *be a minimal prime of* IS . *Then*

$$\dim S_Q \leq \text{pd}_R M .$$

Remarks. If $R = \mathbb{Z}[X_1, \ldots, X_n]$, a polynomial ring over $\mathbb{Z}$, $M = R/(X_1, \ldots, X_n)R$, then this statement reduces at once to the Krull height theorem.

Peskine-Szpiro proved that (4.3) implies both (4.2) and (4.1) in their beautiful thesis [PS_2], and settled the char. $p > 0$ and many other important cases. (4.3)<=>(4.4) is observed in [H_1], where it is also shown that big C-M modules can be used to prove (4.4).

In $[PS_3]$, [Ro], $[H_6]$ it is shown that if R contains a field:

(4.5) Conjecture (new intersection conjecture). Let $0 \to F_d \to \ldots \to F_0 \to 0$ be a finite complex of finitely generated free modules over a local ring (R,m) , whose homology is not zero but is killed by a power of m . Then Krull dim $R \leq d$.

Remark. (4.5) =>(4.4) is easy.

(4.6) Conjecture (direct summand conjecture). Let R be a regular Noetherian ring and let S be a module-finite R-algebra. Then R is a direct summand of S as an R-module.

(4.7) Conjecture (monomial conjecture). Let S be a local ring and $x_1, \ldots, x_n$ a system of parameters. Then for every positive integer t ,

$$x_1^t \ldots x_n^t \notin (x_1^{t+1}, \ldots, x_n^{t+1})S .$$

Remark. (4.7) and (4.8) are studied in $[H_2]$, $[H_8]$, and are equivalent. (4.7) may be reduced to the case where R is complete, local, and unramified.

(4.8) Conjecture (Eisenbud-Evans principal ideal conjecture). Let (R,m) be a local domain, E a finitely generated torsion-free R-module, and let r be the torsion-free rank of E . Let $u \in mE$ and let Q be a minimal prime of Trace (u) where

$$\text{Trace } u = \{\phi(u) : \phi \in \mathrm{Hom}_R(E,R)\} .$$

<u>Then</u> $\dim R_Q \leq r$.

See [EE] for more details. In [H_8] it is shown that (4.6) => (4.8).

But the key point is that the existence of big C-M modules implies all of the conjectures (4.1) - (4.8) and hence appears as a central issue.

To sum up: (4.1) - (4.8) [and (4.9) below] are all known for local rings which contain a field and in dimension ≤ 2 , but are open questions in dimension 3 in the case where the ring R does not contain a field. (An example of such an R is $V[[x_1, \ldots, x_n]]/I$ where V is the p-adic integers (the completion of $\mathbb{Z}_{(p)} = \{m/q : m \in \mathbb{Z} , q \in \mathbb{Z} - p\mathbb{Z}\})$, $x_1, \ldots, x_n$ are formal power series indeterminates, and I is an ideal with $p \notin I$. More specifically, we might have $R = \mathbb{Z}[[x_1, x_2, x_3, x_4]]/I$ and $I = (px_3 - x_1x_2, x_1x_4-x_2x_3, px_4-x_2^2)$.

All these are easily proved from the existence of big C-M modules (although most were first proved in the main cases by other techniques).

Small C-M modules do not seem to give that much more information, but would yield some small progress on Serre's conjecture on multiplicities: we refer the reader to [E], [H_5], [H_7], and [S_2] for more details.

Finally, we want to mention a conjecture, possibly first due in some form to Graham Evans in 1972, but reformulated by the present author as follows:

(4.9) Conjecture (canonical element conjecture). *Let* (R,m) *be a local ring and let*

$$0 \to M \to F_{d-1} \to \dots \to F_o \to K \to 0$$

be an exact sequence, where $K = R/m$, F_i *is a finitely generated free module,* $0 \leq i \leq d-1$, *and* $d = \text{Krull dim } R$.

Let $\mathcal{K}$ *be the Koszul complex on a system of parameters* $x_1, \dots, x_d$ *for* R , *and consider a map of complexes*

$$\begin{array}{ccccccccccccccc}
0 & \to & M & \to & F_{d-1} & \to & \dots & \to & F_1 & \to & F_0 & \to & K & \to & 0 \\
 & & \uparrow{\scriptstyle\phi} & & \uparrow & & & & \uparrow & & \uparrow & & \uparrow & & \\
0 & \to & \mathcal{K}_d & \to & \mathcal{K}_{d-1} & \to & \dots & \to & \mathcal{K}_1 & \to & \mathcal{K}_0 & \to & R/(x_1,\dots,x_n) & \to & 0 \\
 & & \| & & & & & & & & \| & & & & \\
 & & R & & & & & & & & R & & & &
\end{array}$$

(it is always possible to construct such a map). Then independent of the choices of $x_1, \dots, x_n$ *and the map of complexes,* $\phi \neq 0$.

For more information about the Koszul complex, we refer the reader to [M] or [H_5].

This has a tremendous number of equivalent forms: See $[H_8]$. (4.9) is also a consequence of the existence of big C-M modules (hence true if R contains a field), but not known in general. (4.1) - (4.8) <u>all</u> can be deduced from (4.9), which means that knowing (4.9) is almost as good as knowing the existence of big C-M modules. It would not be surprising if (4.9) is equivalent to the existence of big C-M modules.

Remark. Let u be the image in M under ϕ of a generator of K_d . (4.9) is then equivalent to the assertion:

(4.9°) *For every positive integer* t ,

$$x_1^t \cdots x_d^t u \notin (x_1^{t+1}, \ldots, x_d^{t+1})M .$$

[(4.9°) only *seems* stronger.]

(4.9°) is independent of the choice of $x_1, \ldots, x_d$ and also of the choice of the map of complexes.

5. Descent of small Cohen-Macaulay modules.

The results below explain in part why the existence of small Cohen-Macaulay modules in the complete case is crucial.

A local ring (S,n) is called an *approximation* ring if for every finite system of polynomial equations (in finitely many variables) over S , the solutions in S are $\hat{n}$-adically dense in the solutions in $\hat{S}$, where $\hat{n} = n\hat{S}$ is the maximal ideal of the completion $\hat{S}$.

Let R^h denote the Henselization of R (see [Ra], [N], or [H_5]). We recall the crucial results of M. Artin [Ar_1], [Ar_2].

(5.1) Theorem (M. Artin). *Suppose either that* (S,n) *is*

1) *the Henselization of a local ring of a finitely generated algebra over a field or excellent discrete valuation ring, or*

2) *an analytic local ring, i.e., for some* m *, a homomorphic image of the convergent power series ring* $\mathbb{C}\{x_1,\ldots,x_m\}$ *, where* $\mathbb{C}$ *is the complex numbers.*

Then S *is an approximation ring.*

Using the techniques of Peskine-Szpiro [PS_2], pp. 352-254, we can easily show:

(5.2) Theorem. *Let* (R,m) *be a homomorphic image of a regular local ring* (S,n) *which is an approximation ring.*

Suppose that $\hat{R}$ *has a small Cohen-Macaulay module. Then* R *has a small Cohen-Macaulay module.*

Proof. Say $R = S/I$, where $I = (u_1,\ldots,u_h)S$, $u_j \in S$. Let M be a small Cohen-Macaulay module for $\hat{R} \cong \hat{S}/I\hat{S}$, and let

$$0 \to S^{b_d} \overset{A_d}{\to} \cdots \to S^{b_1} \overset{A_1}{\to} S^{b_0} \to M \to 0$$

be a minimal free resolution of M over $\hat{S}$. Then $d = pd_{\hat{S}} M = \dim \hat{S} - depth_{\hat{S}} M = \dim \hat{S} - depth_{\hat{R}} M = \dim \hat{S} - \dim \hat{R} = \dim S - \dim R$. Here, A_i is the b_i by b_{i-1} matrix (over $\hat{S}$) of the map $\hat{S}^{b_i} \to \hat{S}^{b_{i-1}}$. Since $IM = 0$, for each generator $e_i = (0,\ldots,1,\ldots,0)$ of $\hat{S}^{b_0}$, $1 \leq i \leq b_0$, $u_j e_i$ is an $\hat{S}$-linear combination of the rows of A_1 ; i.e., there is a b_0 by b_1 matrix B over $\hat{S}$ such that

$$(*) \qquad (\delta_{ij} u_j) = BA ,$$

and, since the resolution is a complex, we have, of course, that

$$(**) \qquad A_i A_{i-1} = 0 , \quad 2 \leq i \leq d ,$$

as well. Viewing the entries of B and the matrices $A_1,\ldots,A_d$ as unknowns, we see that for each $N \geq 1$ there is a solution ${}_NB$, ${}_NA_i$ for the equations (*), (**) in S such that

$$\left\{\begin{array}{l} {}_NB \equiv B \\ \\ {}_NA_i \equiv A_i \text{ , all } i \end{array}\right. \qquad \text{modulo } \hat{n}^N .$$

By [PS$_2$], Lemma (6.4), for all sufficiently large N the complex

$$0 \to S^{b_d} \overset{{}_NA_d}{\to} \cdots \to S^{b_1} \overset{{}_NA_1}{\to} S^{b_0}$$

has finite length homology (at $S^{b_1},\ldots,S^{b_d}$) and hence, by [PS_2], Cor. (1.9), is acyclic. Hence, if we let ${}_NM = \text{Coker}\ ({}_NA_1)$ we see that $\text{pd}_S\ ({}_NM) = d$ for large N. (The matrices ${}_NA_i$ will have entries in $\hat{n}$, since the A_i do, and so in n, and the resolution will be minimal.) The equation (*) implies that $I({}_NM) = 0$, so that ${}_NM$ may be regarded as an $R = (S/I)$-module, and $\text{depth}_R({}_NM) = \text{depth}_S({}_NM)$ $= \dim S - \text{pd}_S({}_NM) = \dim S - d = \dim R$.

Thus, ${}_NM$ is a small Cohen-Macaulay R-module for all sufficiently large N. QED.

(5.3) Corollary. _Let_ R _be an analytic local ring. If_ $\hat{R}$ _has a small Cohen-Macaulay module, then so does_ R.

Proof. $R = S/I$, where $S = \mathbb{C}\ \{x_1,\ldots,x_r\}$ is an approximation ring which is regular. QED.

(5.4) Corollary. _Let_ (R,m) _be a local ring of a ring_ S _finitely generated over a field or excellent discrete valuation ring._

Suppose that $\hat{R}$ _has a small C-M module. Then_ R^h _has a small C-M module, and hence, so does some pointed étale extension of_ R.

Proof. View S as a homomorphic image of T, where $T = K[X_1,\ldots,X_n]$, a polynomial ring or discrete valuation ring. Then R is a homomorphic image of T_Q for a suitable prime Q and R^h is a homomorphic image of $(T_Q)^h$, which is a regular approximation ring. Thus, R^h has a small C-M module, E , by Theorem (5.2).

But R^h is a direct limit of pointed étale extensions of R and faithfully flat maps. If A is a finite matrix whose cokernel is E , we can choose a pointed étale extension R* of R such that the entries of A lie in R* . Viewing A as a map of free R*-modules, if E* = Coker A we have $E = E^* \otimes_{R^*} R^h$. Since E is C-M , so is E* . Q.E.D.

6. Non-existence of Noetherian Cohen-Macaulay algebras.

The reader may have wondered at the restriction to algebras without unit in Theorem (3.4). It is natural to ask whether a local ring R has a big C-M module which is a (commutative) unital R-algebra. The author does not believe that such algebras exist in general.

If one asks the same question but requires, in addition, that the big C-M module be a <u>Noetherian</u> <u>ring</u>, then one can prove that such modules do not exist. The following result is an improvement of a joint result of W. Heinzer and the author (Proposition (5.9) of $[H_6]$ - but the proof given in $[H_6]$ is not quite correct).

(6.1) Theorem. <u>Let</u> (R,m) <u>be a local ring containing the rationals which is analytically normal, i.e.,</u> $\hat{R}$ <u>is normal. Then</u> R <u>possesses a big C-M module</u> S <u>which is a</u>

Noetherian ring and an R-algebra if and only if R *is itself C-M.*

Remark. Many examples of complete normal local rings containing Q which are not C-M are known: see [H_1 : p. 149], for example. Any such fails to have a Noetherian algebra which is a big C-M module.

The key to the proof of (6.1) is the following lemma, which we prove first:

(6.2) Lemma. *Let* (R,m) *be an analytically normal local ring such that* R *contains the rationals, let* S *be a Noetherian R-algebra, and suppose that*

$$\text{ht } mS = \text{ht } m .$$

Then there exists a commutative diagram of ring homomorphisms

$$\begin{array}{ccc} R & \longrightarrow & S \\ {\scriptstyle\alpha}\downarrow & & \downarrow{\scriptstyle\beta} \\ R' & \underset{\gamma}{\hookrightarrow} & S' \end{array}$$

where α *is faithfully flat, and* R' *is a direct summand of* S' *as an* R'*-module.*

Proof. Let $x_1,\ldots,x_n$ be a system of parameters for R . Let Q be a minimal prime of $(x_1,\ldots,x_n)S$. Let $S' = \hat{S}_Q/q$

for a minimal prime q of S_Q such that $\dim S_Q/q = n$, so that the images of $x_1,\ldots,x_n$ in S' are a system of parameters. Let $K \subset \hat{R}$ be a copy of the residue class field, identify K with $\mathrm{Im}(K \to S')$, and let $L \subset S'$ be a copy of the residue class field with $K \subset L$. Let $A = K[[x_1,\ldots x_n]] \subset \hat{R}$, so that $\hat{R}$ is module-finite over the regular ring A, let $B = L[[x_1,\ldots,x_n]] \supset A$, and let $R' = B \otimes_A \hat{R}$. Since B is A-flat, R' is $\hat{R}$-flat. Since $\hat{R}$ is a finite module over A, R' is a finite module over B. Since the images of $x_1,\ldots,x_n$ in S' are a system of parameters, we have a natural map $B \to S'$ which is injective, and S' is a finite module over B. Since we have $B \hookrightarrow S'$ and $\hat{R} \to S'$ as A-algebras, we have a map $B \otimes_A \hat{R} = R' \to S'$. Since R' is module-finite over B and $B \to B \otimes \hat{R}$ is injective (even $B \to B \otimes \hat{R} \to S'$ is injective from B), we know $\dim R' = n$.

We next show that R' is a domain. Let R, G, H denote the fraction fields of A, B, and $\hat{R}$, respectively. Since B is A-flat, the injection $\hat{R} \hookrightarrow H$ yields an injection

$$R' = B \otimes_A \hat{R} \hookrightarrow B \otimes_A H$$

and since $H = \hat{R} \otimes_A F$, the above map is simply $R' \to R \otimes_A F$, and so we have

$$R' \hookrightarrow R' \otimes_A F \cong (B \otimes_A \hat{R}) \otimes_A (R \otimes_A F)$$

$$\cong (F \otimes_A B) \otimes_A (\hat{R} \otimes_A F)$$

$$\cong (F \otimes_A B) \otimes_F H$$

Since we have $F \otimes_A B \hookrightarrow G$ and F is a field, we have an injection

$$(F \otimes_A B) \otimes_F H \to G \otimes_F H .$$

Thus, it will suffice to show that $G \otimes_F H$ is a domain. Let $\bar{F}$ be the algebraic closure of R in G. By [Z_1, pg. 198, Corollary 2] it suffices to show that $\bar{F} \otimes_F H$ is a domain.

Now, any element of $\bar{F}$, after multiplication by a suitable element of A, becomes integral over A, and hence lies in B, not just in G. Moreover, a power series in B integral over A must have all its coefficients in a single finite (over K) algebraic extension K_1, $K \subset K_1 \subset L$. [To see this we may suppose L is the algebraic closure of K (recall: $B = L[[x_1,\ldots,x_n]]$). The power series u has only finitely many conjugates (lying among the roots of its monic polynomial) in G/F. Let G_0 be the (finite) field extension of F generated by these and let K^* be the subfield of K generated by their coefficients. Each automorphism of L/K induces an automorphism of B over A (acting on the coefficients) and hence fixes K^*. Moreover, by acting on coefficients Gal (K^*/K) is <u>embedded</u> in Gal (G_0/F), whence K^* is a finite field extension of K.]

If K^* is any finite field extension of K in L, let $B_{K^*} = F \otimes_A K^*[[x_1,\ldots,x_n]]$. Then

$$\overline{F} = \bigcup_{K^*} B_{K^*} ,$$

whence it suffices to show that when L is finite over K (and then $\overline{F} = G$),

$$G \otimes_F H$$

is a domain. But when L is finite over K, $B = L \otimes_K A$ is finite over A, $G \otimes_F H$ is a localization at a multiplicative system of $L \otimes_K H$, and to show that $L \otimes_K H$ is a domain it suffices to show that K is algebraically closed in H. Any element of H integral over K is integral over $\hat{R}$ and, hence, in $\hat{R}$, and K is integrally closed in $\hat{R}$ (the integral closure of a field in a domain is a field, and K is a maximal subfield of $\hat{R}$).

Thus, R' is a <u>domain</u> module-finite over B and hence a local domain of dimension n. Since S' is module-finite over the image of R' and $\dim S' = n$, $R' \to S'$ is injective.

If we write $\hat{R} = K[[U_1,\ldots,U_t]]/P$, where $P = (f_1,\ldots,f_s)$ has height $r = t-n$, then the ideal of r size minors $J = I_r(\overline{\partial f_i/\partial U_j})$, where $\overline{}$ is reduction modulo P, defines the singular locus in $\hat{R}$, i.e., $\hat{R}_Q$ is not regular $\Leftrightarrow Q \supset J$, by [N, Thm. (46.3)]. Let $R'' = L[[U_1,\ldots,U_t]]/(f_1,\ldots,f_s) \cong$ the completion of $L \otimes_K \hat{R}$ at $L \otimes_K m$. We have a map $R'' \to R'$ which is clearly surjective. Moreover, the singular locus in R'' is defined by JR'' and R'' is flat over $\hat{R}$, so that depth $JR'' \geq 2$ again. It follows that R'' is normal and $R'' \to R'$ is an isomorphism, since the map is surjective and $\dim R'' = \dim R'$.

Since $R' \to S'$ is a map of complete local domains with the same residue class field and the system of parameters $x_1,\ldots,x_n$ for R' maps to a system of parameters for S' , S' is a finite-module over R' . If d is the degree of the field extension, $(1/d)\mathrm{Tr}$ retracts S' onto R' (Tr is the field trace). Cf. [H_2]. Q.E.D.

Proof of Theorem (6.1). "If" is trivial: let $S = R$. To prove "only if", suppose S is a big C-M module for R and a Noetherian R-algebra, and let $x_1,\ldots,x_n$ be a s.o.p. for R which is an S-sequence. Then $x_1,\ldots,x_n$ is an S-sequence => height $(x_1,\ldots,x_n)S = n$ => height $mS = n$, and we may apply the lemma. Let α, β, γ, R', S' be as in Lemma 6.2.

We shall show that $x_1,\ldots,x_n$ is an R-sequence. Suppose that

$$rx_{t+1} = r_1x_1 + \ldots + r_tx_t ,$$

where $r_1,\ldots,r_t$, $r \in R$. We must show that $r \in (x_1,\ldots,x_t)R$.

Since $x_1, \ldots, x_t$ is an S-sequence, we know that $r \in (x_1,\ldots,x_t)S \Rightarrow r \in (x_1,\ldots,x_t)S' \Rightarrow r \in (x_1,\ldots,x_t)S' \cap R$. But since R' is a direct summand of S', every ideal is contracted, and since α is faithfully flat, every ideal of R is contracted. It follows that every ideal of R is contracted from S', and so $r \in (x_1,\ldots,x_t)S' \cap R = (x_1,\ldots,x_t)R$. Q.E.D.

References

[Ar_1] M. Artin, On the solutions of analytic equations, Invent. Math. 5 (1968), 277-291.

[Ar_2] M. Artin, Algebraic approximation of structures over complete local rings, Publ. Math. I.H.E.S., Paris, No. 36, 1969.

[Au_1] M. Auslander, Modules over unramified regular local rings, Illinois J. of Math. 5 (1961), 631-645.

[Au_2] M. Auslander, Modules over unramified regular local rings, Proc. Intern. Congress of Math., 1962, 230-233.

[AB_1] M. Auslander and D. A. Buchsbaum, Homological dimension in Noetherian rings, Proc. Nat. Acad. Sci. USA 42 (1956), 36-38.

[AB_2] M. Auslander and D. A. Buchsbaum, Homological dimension in local rings, Trans. Amer. Math. Soc. 85 (1957), 390-405.

[AB_3] M. Auslander and D. A. Buchsbaum, Codimension and multiplicity, Ann. of Math. 68 (1958), 625-657; corrections, Ann. of Math. 70 (1959), 395-397.

[B] H. Bass, On the ubiquity of Gorenstein rings, Math. Z. 82 (1963), 8-28.

[E] D. Eisenbud, Some directions of recent progress in commutative algebra, Proc. of Symposium in Pure Math. 29, Amer. Math. Soc., 1975, 111-128.

[EE] D. Eisenbud and E. G. Evans, A generalized principal ideal theorem, Nagoya Math. J. 62 (1976), 41-53.

[FR] D. Ferrand and M. Raynaud, Fibres formelles d'un anneau local noethérien, Ann. Sci. Ec. Norm. Sup. (4) 3 (1970), 295-311.

[FFGR] R. Fossum, H.-B. Foxby, P. Griffith, and I. Reiten, Minimal injective resolutions with applications to dualizing modules and Gorenstein modules, Publ. Math. I.H.E.S. Paris No. 45, 1975, 193-215.

[F_1] H.-B. Foxby, On the μ^i in a minimal injective resolution, Math. Scand. 29 (1971), 175-186.

[F_2] H.-B. Foxby, Applications of isomorphisms between complexes, preprint (Copenhagen University).

[F_3] H.-B. Foxby, On the μ^i in a minimal injective resolution II, Copenhagen Math. Inst. preprint series, Aug. 1976, No. 20.

[Gr] P. Griffith, A representation theorem for complete local rings, J. Pure and Applied Algebra 7 (1976), 303-315.

[GH] A. Grothendieck (notes by R. Hartshorne), "Local Cohomology", Lecture Notes in Math. no. 41, Springer, New York, 1967.

[H_1] M. Hochster, Cohen-Macaulay modules, Proc. Kansas Commutative Algebra Conference, Springer-Verlag Lecture Notes in Math., No. 311, New York, 1973 120-152.

[H_2] M. Hochster, Contracted ideals from integral extensions of regular rings, Nagoya Math. J. 51 (1973), 25-43.

[H_3] M. Hochster, Deep local rings, preliminary preprint, Århus University preprint series, December, 1973.

[H_4] M. Hochster, The equicharacteristic case of some homological conjectures on local rings, Bull. Amer. Math. Soc. 80 (1974), 683-686.

[H_5] M. Hochster, "Topics in the homological theory of modules over commutative rings", C.B.M.S. Regional Conference Series in Math. No. 24, Amer. Math. Soc., Providence, 1975.

[H_6] M. Hochster, Big Cohen-Macaulay modules and algebras and embeddability in rings of Witt vectors, "Proc. of the Queen's University Commutative Algebra Conference" (Kingston, Ontario, Canada, 1975) Queen's Papers in Pure and Applied Math. No. 42, 106-195.

[H_7] M. Hochster, Some applications of the Frobenius in characteristic zero, Bull. Amer. Math. Soc., to appear.

[H_8] M. Hochster, The canonical element conjecture, in preparation.

[Iv] B. Iversen, Amplitude inequalities for complexes, Aarhus University Preprint Series No. 36 (1976/77).

[K] I. Kaplansky, "Commutative Rings", Allyn and Bacon, Boston, 1970. Revised ed., 1974.

[LV] G. Levin and W. Vasconcelos, Homological dimensions and Macaulay rings, Pacific J. Math. 25 (1968), 315-328.

[M] H. Matsumura, "Commutative algebra", Benjamin, New York, 1970.

[N] M. Nagata, "Local rings", Interscience, New York, 1962.

[PS_1] C. Peskine and L. Szpiro, Notes sur un air de H. Bass, unpublished preprint (Brandeis University, Waltham, Massachusetts).

[PS_2] C. Peskine and L. Szpiro, Dimension projective finie et cohomologie locale, Publ. Math. I.H.E.S., Paris, No. 42, 1973, 323-395.

[PS_3] C. Peskine and L. Szpiro, Syzgies et multiplicités, C. R. Acad. Sci., Paris, Ser. A. (1974), 1421-1424.

[Ra] M. Raynaud, "Anneaux locaux henséliens", Springer-Verlag Lecture Notes in Math. No. 169, New York, 1970.

[Re] D. Rees, The grade of an ideal or module, Proc. Cambridge Philos. Soc. 53 (1957), 28-42.

[Ro] P. Roberts, Two applications of dualizing complexes over local rings, Ann. Scient. Ec. Norm. Sup. (4) 9 (1976), 103-106.

[S_1] J.-P. Serre, Sur la dimension homologique des anneaux et des modules Noethériens, Proc. Internat. Sympos. Algebraic Number Theory, Tokyo, 1955, 175-189.

[S_2] J.-P. Serre, Algèbre locale. Multiplicités. Springer-Verlag Lecture Notes in Math. No. 11, New York, 1965.

[ZS] O. Zariski and P. Samuel, "Commutative algebra", Vol. 1, Princeton, Van Nostrand, 1958.

Rings of Bounded Module Type

Roger Wiegand[1]
University of Nebraska
Lincoln, Nebraska 68588

A ring A is an FGC-ring provided every finitely generated left A-module is a direct sum of cyclic modules. In 1976 a very satisfactory structure theorem was proved for the commutative FGC-rings. (Many mathematicians contributed to the final solution; see [5] for a self-contained exposition of the proof.) Any attempt to characterize non-commutative FGC-rings runs into serious difficulties unless some restrictions are imposed. For one thing there are rings that are FGC-rings for the rather silly reason that all of their finitely generated modules are cyclic. To avoid such pathology we will deal only with rings that are finitely generated as modules over their centers. We will fix the following notation for the rest of this paper: R *is a commutative ring, and* A *is an* R-algebra *that is finitely generated as an* R-module.

The problem we will consider first is this: What conditions are forced on R by the assumption that A is an FGC-ring? Clearly we will have to impose some sort of non-degeneracy condition, for example, that R is embedded in A. In order to have a basis for conjecture, we will state the structure theorem for commutative FGC-rings. (The form given here is easily deduced from the results in [5], and conversely.)

[1]Research for this paper was supported in part by a grant from the National Science Foundation.

<u>Theorem</u> 1: The commutative ring R is an FGC-ring if and only if it satisfies the following four conditions:

(1) Each ideal of R has only finitely many minimal primes.

(2) R_M is an almost maximal valuation ring for each maximal ideal M.

(3) If I is an ideal with a unique minimal prime P, the ideals between I and P form a chain.

(4) Every finitely generated ideal of R is principal.

Suppose $R \subseteq A$ and A is an FGC-ring. Which of the properties (1)-(4) are valid? Property (1) is very likely true, and I have a proof for countable rings. Property (2) is certainly false. For example, take $R = k[t^2,t^3]$, $A = k[t]$, where k is a field. Similarly (4) is false, for if R is any Dedekind domain, then every finitely generated R-module is a direct sum of 2-generator modules. It follows that the ring of 2×2 matrices over R is an FGC-ring.

This last observation brings us to the rings of the title. We let (B_n) denote the condition (on A) that every finitely generated left A-module is a direct sum of modules generated by at most n elements. We will say A has <u>bounded module type</u> provided A has (B_n) for some $n \geq 1$. Clearly, a ring has (B_n) if and only if its $n\times n$ matrix ring is an FGC-ring. Keeping the examples above in mind, we make the following

<u>Conjecture</u>: Let A be an FGC-ring, finitely generated as a module over its center. Then there is a ring S, of bounded module type and contained in the center of A, such that A is a finitely generated S-module.

We remark that $k[t^2, t^3]$ does not have bounded module type, in view of the following result:

Theorem 2 (Warfield, [3]): If R is a commutative ring of bounded module type, then R_M is a valuation ring for every maximal ideal M.

In fact, Warfield proves that if R is local but not a valuation ring, then for each **n** there is an indecomposable module requiring exactly **n generators.** In the same paper he shows that a commutative Noetherian ring of bounded module type satisfies (B_2). It is unknown whether the word "Noetherian" can be dropped; I will return to this question after recording my only definitive result on noncommutative FGC-rings.

Theorem 3: Let R be a countable commutative ring, and let A be an R-algebra that is finitely generated as an R-module. Assume that the kernel of $R \to A$ is a nil ideal. If A has bounded module type, then every ideal of R has only finitely many minimal primes.

Proof: By passing to a suitable matrix ring, we may assume A has FGC. If I is an ideal of R, all the hypotheses carry over to $R/I \to A/IA$. (This is the reason we didn't make the more natural assumption that $R \subseteq A$.) Thus it will suffice to prove that R has only finitely many minimal primes. Moreover, we may assume R is reduced.

Suppose, then, that R has infinitely many minimal primes. It is well known that R must contain an infinite direct sum of ideals. (One way to prove this is to note that the minimal prime spectrum of R is an infinite regular Hausdorff space, so it must contain an infinite family of pairwise disjoint open sets.) Thus, let $\oplus \sum_n I_n \subseteq R$, where each I_n is a non-zero ideal of R. Let $\{(a_n,b_n)\}$ be a list of the elements of $A \oplus A$, and let ϕ be the $\omega \times 2$ matrix whose nth row is (a_n,b_n). We define a left A-module M by the exact sequence

$$\oplus \sum_n I_n A \xrightarrow{\phi} A \oplus A \xrightarrow{\rho} M \to 0 \qquad (*)$$

where ϕ denotes right multiplication by the matrix. We will show that M is not a direct sum of cyclic modules, the desired contradiction.

Suppose $M = Ax_1 \oplus \ldots \oplus Ax_t$, and set $x = x_1 + \ldots + x_t$, say, $x = (a_n,b_n)\rho$. Since $I_n \neq 0$ we can find a prime ideal $P \not\supseteq I_n$. Then $I_nA_P = A_P$, and when we localize (*) at P we see that (a_n,b_n) is in the image of ϕ_P. Therefore x maps to 0 in M_P, and it follows that $M_P = 0$, that is, ϕ_P is onto. If m is any index different from n, we have $I_m \cap I_n = 0$, so $I_m \subseteq P$. Therefore I_mA_P is contained in J, the Jacobson radical of A_P. Thus we have a surjection $\oplus \sum_{m \neq n} K_m \oplus A_P \xrightarrow{\phi_P} A_P \oplus A_P$, where each $K_m \subseteq J$. Since ϕ_P is defined by elements of A_P, ϕ_P must carry $\oplus \sum_{m \neq n} K_m$ into $J(A_P \oplus A_P)$. But then by Nakayama's lemma ϕ_P maps A_P onto $A_P \oplus A_P$, a contradiction.

Corollary: Let R be a countable commutative ring of bounded module type. Then every ideal of R has only finitely many minimal primes.

Using an entirely different (and much more sophisticated) method, B. Midgarden, a student at the University of Nebraska, has proved that the word "countable" may be removed from the corollary. Thus commutative rings of bounded module type satisfy (1) of Theorem 1, and by Warfield's theorem part of Condition (2). As for almost maximality, I haven't a clue. Condition (3) seems unlikely to be true for rings of bounded module type. The rings constructed by S. Wiegand in [6] might provide examples satisfying (B_{n+1}) but not (B_n); these rings have homomorphic images not satisfying Condition (3). Here is an amusing result in the other direction.

Theorem 4: Let R be a commutative ring satisfying Conditions (1), (2) and (3) of Theorem 1. Then R satisfies (B_2).

Proof: Let M be a finitely generated R-module, and let I be its annihilator. I claim there is a direct summand D_1 of M, generated by two elements and having I as its annihilator. This will complete the proof. For, we can repeat the argument on the factor M/D_1, and continue the process, to get an ascending chain of direct summands $D_1 \subset D_1 \oplus D_2 \subset D_1 \oplus D_2 \oplus D_3 \subset \ldots$, such that $(0:D_k) \subsetneqq (0:D_{k+1}) = (D_1 \oplus \ldots \oplus D_k : M)$. If M is generated by n elements, the whole process stops after at most n steps. (If not, we would get a contradiction by localizing the summand $D_1 \oplus \ldots \oplus D_{n+1}$ at a maximal ideal containing the annihilator of D_{n+1}.)

To prove the claim, we may assume M is faithful, since properties (1), (2), (3) carry over to homomorphic images. Moreover, we may assume R has a unique minimal prime P. (The minimal primes of R are pairwise comaximal, by (2), so one can apply the Chinese remainder theorem and then lift idempotents.) By [5, Lemma 17] there is at most one maximal ideal M for which $P_M \neq 0$. (If $P = 0$ let M be an arbitrary maximal ideal.) It follows that the set of zero-divisors of R is contained in M, so K, the classical quotient ring of R, is a localization of R_M. Then, by results of Gill, Klatt and Levy, [5, Lemmas 4 and 6], K is a self-injective ring and hence an injective R-module.

As in §7 of [5] we can find an element $x \in M$ with annihilator 0. Let $\theta: M \to K$ extend the embedding $Rx \to K$ taking x to 1. Then the image of θ is isomorphic to a finitely generated ideal H containing a non-zerodivisor. Certainly H is projective, being locally free of rank 1. The proof will be complete once we show that H is generated by two elements. A direct proof is not hard, but the quickest approach is to notice that every non-minimal prime ideal of R is contained in a unique maximal ideal (by the proof of Lemma 10 of [5]); therefore the maximal ideal space of R has dimension 0 or 1. Also, by results of Heinzer, Ohm and Pendleton, [1, Proposition 2.2], the maximal ideal space is Noetherian, so Swan's theorem [2] implies that H may be generated by two elements.

The argument at the beginning of the proof was really a special case of a more general result. It turns out that M has the ascending chain condition on direct summands (even if the annihilators are not nested). In fact this is true for any commutative ring satisfying Condition (1), since the maximal ideal space is then Noetherian (proof of [1, Proposition 2.2]), and we can appeal to the following:

Theorem 5: Let R be a commutative ring with Noetherian maximal ideal space. Then every finitely generated module has the ascending chain condition on direct summands.

Proof: Let M be a finitely generated R-module containing an infinite direct sum $A_1 \oplus A_2 \oplus A_3 \oplus \ldots$, where $A_i \oplus B_i = M$. Let X be the j-spectrum of R, that is, the set of prime ideals that are intersections of maximal ideals. Then X is compact in the patch topology, [4, §1], and Noetherian in the Zariski topology. Set $X_k = \{P \in X \mid \mu(R_P, M_P) = k\}$, where μ = number of generators required. These sets X_k are open and closed in the patch topology on X.

Suppose $P \in X_k$. Then $(A_i)_P = 0$ if and only if $\mu(R_P, (B_i)_P) \geq k$, so if we set $Z_i = \{P \in X \mid (A_i)_P = 0\}$, we see that $Z_i \cap X_k$ is closed in the Zariski topology on X_k. Thus, for each n, the set $V_n = (\bigcap_{i \geq n} Z_i) \cap X_k$ is closed in the Zariski topology on X_k, and therefore open in the patch topology. For any prime P we have $(A_i)_P = 0$ for large i, that is, the sets V_i cover X_k. By compactness of the patch topology, we have $X_k = V_n$ for some integer n. Since there are only finitely many sets X_k to deal with, we obtain an integer N such that $X \subseteq \bigcap_{i \geq N} Z_i$, that is, $A_i = 0$ for each $i \geq N$.

Corollary: Let R be a countable commutative ring and let A be an R-algebra, finitely generated as an R-module. If A has bounded module type then A has no infinite family of orthogonal idempotents, and every finitely generated left A-module is a direct sum of indecomposable modules.

If A is commutative, the word "countable" may be dropped, in view of Midgarden's work mentioned earlier. It would be interesting to know whether the conclusions of the corollary are true for commutative rings satisfying the following weaker property: (*) There is an integer n such that every finitely generated indecomposable module is generated by n elements. If so, then (*) would of course be equivalent to having bounded module type.

References

[1] T. S. Shores and R. Wiegand, "Rings whose finitely generated modules are direct sums of cyclics", J. Algebra 32 (1974), 152-172.

[2] R. G. Swan, "The number of generators of a module", Math. Z. 102 (1967), 318-322.

[3] R. B. Warfield, Jr., "Decomposability of finitely presented modules", Proc. Amer. Math. Soc. 25 (1970), 167-172.

[4] R. Wiegand, "Dimension functions on the prime spectrum", Comm. in Algebra 3 (1975), 459-480.

[5] R. Wiegand and S. Wiegand, "Commutative rings whose finitely generated modules are direct sums of cyclics", Lecture Notes in Mathematics 616, Springer (1977), 406-423.

[6] S. Wiegand, "Locally maximal Bezout domains", Proc. Amer. Math. Soc. 47 (1975), 10-14.

INJECTIVE QUOTIENT RINGS OF COMMUTATIVE RINGS

Carl Faith[1]

Rutgers, The State University
New Brunswick, N. J. 08903
and
The Institute for Advanced Study
Princeton, N. J. 08540

INTRODUCTION

In the broadest sense, this is a study of commutative rings which satisfy the <u>(finitely) pseudo-Frobenius</u> (or (F)PF) condition: All (finitely generated) faithful modules generate the category mod-R of all R-modules. These rings include: Prüfer rings, almost maximal valuation rings, self-injective rings, e.g., quasi-Frobenius (QF) and pseudo-Frobenius (PF) rings, and finite products of these. (In fact, any product of commutative FPF rings is FPF [34]; hence, any product of commutative PF rings is FPF (cf. §9).)

If R is FPF, so is its (classical) ring of quotients $Q_{c\ell}(R)$ and its maximal quotient ring $Q_{max}(R)$. All known FPF rings are <u>(classically) quotient-injective</u> in the sense that $Q_{c\ell}$ is injective.[2] We conjecture that all FPF rings are quotient-injective, and prove this in the three cases: (1) local rings (Proposition 7 and Theorem 9B); (2) Noetherian rings (Theorem 11; Endo's Theorem [25]); (3) reduced rings (Proposition 3B and Theorem 4). Moreover, any FPF commutative ring R splits, $R = R_1 \times R_2$, where R_1 is semihereditary, and R_2 has essential nilradical. (If R is semilocal or Noetherian, then R_2 is injective.) Thus any reduced FPF ring has regular injective $Q_{c\ell}$, and conversely any quotient-injective semihereditary ring <u>is</u> FPF (Theorem 4).

A ring is <u>pre-(F)PF</u> iff all (finitely generated) faithful ideals are generators, and we

---

[1] This paper was written while I was a visitor at The Institute for Advanced Study. I wish to thank the faculty for granting me this inestimable privilege. It is also a pleasure to thank Ms. E. Laurent for her many kindnesses and much help.

[2] In general, $Q_{c\ell}$ is injective as an R-module iff it is a self-injective ring [21].

show this occurs iff all such ideals are actually projective. This is proved via a partial converse of Azumaya's theorem (corollary to Proposition 5A) stating that all faithful finitely generated projectives are generators. The partial converse states that all "rank-1" generators are finitely generated projective. (See Theorem 1C and Propositions 1D and 1F.) This enables us to prove that any FPF ring R has flat epic Q_{max} (Theorem 1E). A ring R is right Kasch if every simple right module embeds in R; equivalently, maximal right ideals have nonzero left annihilators. Clearly, any commutative Kasch ring is pre-PF. Moreover, every pre-PF commutative ring has Kasch Q_{max} (Proposition 1G).

Noetherian quotient-injective rings have been characterized by Bass [21]: The zero ideal is unmixed and all of its primary components are irreducible.[3] In the general case, while the problem of characterizing quotient-injective rings is still open, Vámos [19] determined all <u>fractionally self-injective</u> (= <u>FSI</u>) rings, that is, rings such that every factor ring is quotient-injective (see Theorem 19), and related them to the structure of σ-<u>cyclic rings</u>, that is, rings over which every finitely generated module is a direct sum of cyclics. It follows easily from the structure theory of Brandal [27], Vámos [19], and the Wiegands [20] that every σ-cyclic ring is quotient-injective. (See Theorem 19.)

The condition that every factor ring of R is FPF is called <u>CFPF</u>, and is related to Vámos' condition FSI. The truth of our conjecture would imply that R CFPF => FSI. A local ring R is CFPF iff R is an almost maximal valuation ring (Theorem 5B). Thus CFPF <=> FSI for a local ring R by a theorem of Vámos [19]. (These results imply that not every valuation ring (VR) is quotient-injective, since otherwise every factor ring of a VR would be quotient-injective, hence FSI whence almost maximal.) Also CFPF => FSI for Noetherian R (Corollary 12C).

It is shown that a local ring R is FPF iff $Q_{c\ell}$ is injective and the zero divisors P is a "waist" of R such that R/P is a valuation ring. (This general-

[3] Another characterization: The dual of any finitely generated module is reflexive [21].

izes Faith-Zaks [5] for VR's. (See Theorem 9B.))

A local ring R is σ-cyclic iff R is an almost maximal VR by theorems of Kaplansky, Gill, Warfield and others. Since this is equivalent to CFPF, this suggests that FPF is a kind of pre-σ-cyclic condition for local rings. This analogy is made more explicit by Theorem 23 which characterizes an FPF local ring by the condition that every faithful module generated by two elements is σ-cyclic. (An equivalent condition is that every submodule K of R^2 which for all $a \in R$ intersects R^2a in 0 embeds in a direct summand $\approx R$.)

The condition CFP^2F defined analogously to CFPF for finitely presented modules is taken up in Section 4, the main theorem being that this is a local property: R is CFP^2F iff the local ring R_P is CFP^2F for all primes P (Theorem 5D.) Actually, locally FP^2F implies FP^2F (ibid.). While an FP^2F local ring need not be a valuation ring, CFP^2F characterizes valuation rings among local rings (Theorem 5B). Any flat-ideal ring, e.g., any semihereditary ring R is CFP^2F (Theorem 5H).

TABLE OF CONTENTS

1. MAXIMAL QUOTIENT RINGS

We need the concept of a dense submodule M of a right R-module P, namely, $Hom_R(S/M, P) = 0$ for all submodules S of P containing M. This is equivalent to the requirement: For any pair $x, y \in P$, with $x \neq 0$, of the existence of $r \in R$ such that $yr \in P$ and $xr \neq 0$ (e.g., see [3b], p. 79, 19.32.) If R is commutative, and if R is dense in P, then M is a dense submodule of P iff M is faithful. This follows from the trivial Proposion 1A (p. 11), and is given by Lambek as the definition of a dense ideal of R (see [24], p. 37).

It is easy to see that the intersection of two dense submodules is dense, and indeed that the set $D^r(R)$ of dense right ideals in a ring R is a Gabriel filter, that is, defines a Gabriel topology on R (see, e.g., [11], p. 149); we shall have recourse to Gabriel filters again in Theorem B, and the reader is referred to Stenstrom [11], or [3a], Chapter 16.

If P is nonsingular in the sense that no nonzero element annihilates an essential right ideal, then a submodule M is dense iff essential (see [3b], p. 80, 19.32(d)).

Following Findlay-Lambek (see Lambek [24], or [3b], Chapter 19), if M and P are right R-modules, and $P \supseteq M$, then P is a rational extension of M if M is a dense submodule. Thus, a rational extension P of M is an essential extension, hence embeddable in the injective hull $E = E(M)$, and in fact in the unique maximal rational extension $\overline{M}$ of M contained in E. If $B = \text{End}\, M_R$, then $\overline{M} = ann_E ann_B M = \bigcap_{f \in B} \ker f$ and $\ker f \supseteq M$.

The maximal rational extension $\overline{R}$ of R is a ring containing R as a subring, and $\overline{R}$ is denoted $Q^r_{max}(R)$, the (Johnson-Utumi) maximal right quotient ring of R. As Lambek showed, $Q = Q^r_{max}(R)$ is isomorphic to $\text{End}_B E$ under $q \mapsto q(1)$, where $E = E(R)$, $B = \text{End}\, E_R$. (See [24], p. 94, Lemma 1 and Proposition 1 or [3b], Proposition 19.34.) If R is commutative, then so is Q (see [24],

p. 39, Proposition 2) and then Q is injective iff $B = \mathrm{End}\, E_R$ is commutative. (For fun, we prove both of these theorems in Theorem 28 to close out Section 5.)

If R is a right nonsingular ring (= one in which every right annihilator $x^{\perp}$ is essential iff $x = 0$), then one knows that every essential right ideal is dense, and hence that $E = \overline{R} = Q^r_{max}(R)$ is injective, and von Neumann regular ([3b], p. 81, 19. 35). Moreover, there is a 1-1 correspondence between principal right ideals eQ of $Q = Q^r_{max}(R)$ (= right ideals of Q generated by an idempotent = direct summands of Q in mod-Q) and complement right ideals of R given by contraction $eQ \mapsto eQ \cap R$ (ibid., p. 82).

The full right quotient ring $Q^r_{c\ell}(R)$, when it exists, embeds canonically in $Q^r_{max}(R)$. Moreover, $R \hookrightarrow Q^r_{c\ell}(R)$ is an epic in the category RINGS, and $Q^r_{c\ell}(R)$ is flat as a left R-module. In general, a flat epic of a ring R denotes an embedding $R \hookrightarrow T$ of rings for which T is a left flat R-module and $R \hookrightarrow T$ is an epic in RINGS. Every ring R has a maximal flat-epic $R \hookrightarrow Q^r_{tot}(R)$ which is unique up to isomorphism. If R is commutative and Noetherian, then $Q_{c\ell} = Q_{tot} = Q_{max}$. (Consult Stenstrom [11], Example 4, p. 237.) Also, if R is right nonsingular (n. s.), of finite Goldie dimension, then by Goldie's theorems,

$$Q^r_{c\ell} = Q^r_{tot} = Q^r_{max} = E \ .$$

We study conditions on a commutative ring R under which $Q_{tot} = Q_{c\ell}$. Of course, this always happens when $Q_{c\ell}$ is a von Neumann regular ring since the latter class of rings has no proper epics, that is, R von Neumann regular $\Rightarrow R = Q_{c\ell}(R) = Q_{tot}(R)$. This happens, e.g., when R is a Rickart Ring (Lemma 3E).

Cateforis [2] characterized the condition when $R \hookrightarrow Q^r_{max}$ is a right flat epic (i.e., Q^r_{max} is a right flat R-module) for a s.h. ring R by the property that finitely generated nonsingular right R-modules are projective. Goodearl [6] generalized this by the substitutions "n. s." for "s. h." and "embeddable in a projective" for a "projective".

Note, for commutative R, that Cateforis' result can hold (for R s.h.) only if $\overline{R} = Q_{c\ell}$, that is, $R \hookrightarrow Q_{max}$ is a (flat) epic for a commutative s.h. ring R iff $Q_{max} = Q_{c\ell}$ canonically. (See Theorem 4.)

If $\phi : R \to S$ is a ring homomorphism, then the following two theorems determine when ϕ is a (left) flat epic.

The first theorem is essentially that of Silver [10].

Theorem A. *If $\phi : R \to S$ is a ring homomorphism, then the following are equivalent conditions:*

(a) ϕ *is a ring epic (= an epic in the category of rings).*

(b) $S \otimes_R S \hookrightarrow S$ *is an isomorphism.*

(c) mod-$S \hookrightarrow$ mod-R *is full.*

(d) $S \otimes_R (S/\mathrm{im}\phi) = 0$.

Theorem B. (N. Popescu and T. Spircu) *For a ring homomorphism $\phi : R \to S$, the f.a.e.:*

($\overline{a}$) ϕ *is a (left) flat epic, that is, a ring epic and S is a flat left (pull-back) R-module.*

($\overline{b}$) *S is canonically isomorphic to the quotient ring defined with respect to the Gabriel filter F of all ideals I of R such that $\phi(I)S = S$, that is, the isomorphism $\sigma : S \to R_F$ is such that $\sigma\phi : R \to R_F$ is canonical.*

($\overline{c}$) (i) *For every $x \in S$ there exist $r_1, \ldots, r_n \in R$ and $q_1, \ldots, q_n \in S$ such that $x\phi(r_i) \in \phi(R)$ and $\Sigma_{i=1}^n \phi(r_i)q_i = 1$.*

(ii) *If $\phi(I) = 0$, then in (i) the q_i can be picked so that $\phi(I)q_i = 0$, $i = 1, \ldots, n$.*

Note: If $\phi : R \to S$ is inclusion, then (ii) is vacuous.

This theorem and proof is contained in Stenstrom's book [11], p. 227. Also see Storrer [12,13] for related results.

This theorem implies Theorem 1E, which states that for any commutative FPF ring $R \hookrightarrow Q_{max}$ is a flat epic, but this requires an additional lemma in the form of Proposition 1B in Section 2.

Theorem C. *If R is any FPF commutative ring, then any flat epic overring Q of R is FPF, e.g., $Q_{c\ell}$ and Q_{max} are FPF.*

Proof. Let M be any f.g. faithful Q-module, let $m_1, \ldots, m_n$ generate M, and let M_1 be the R-submodule generated by these. Since $R \subseteq Q$, then M_1 is a faithful R-module, hence generates mod-R, so $M_2 = M_1 \otimes_R Q$ generates mod-Q. (To wit: $M_1^n \approx R \oplus X \Rightarrow M_2^n \approx Q \oplus Y$.) Now the kernel K of the canonical epic $M_2 \to M$ is zero since $R \hookrightarrow Q$ is a flat epic, since $M_1 \hookrightarrow M$ as R-modules. Thus $M \approx M_2$ generates mod-Q

Thanks are due to the referee for shortening the proof. Actually, my original proof proved more: Every ring Q between R and Q_{max} is FPF. This also follows from Theorem 1E.

1_{bis} KASCH RINGS

Proposition D. *A ring Q is said to be right Kasch if it satisfies the following equivalent conditions: (K1) Every simple right module embeds in Q; (K2) $\ell_Q M \neq 0$ for every maximal right ideal M; (K3) Q has no dense right ideals $\neq Q$. (K4) E(Q) cogenerates mod-Q. When this is so, then $Q = Q^r_{max}(Q)$.*

Proof. See [11], p. 235, Lemma 5.1.

The next corollary is a well-known consequence.

Theorem E. *A ring R has right Kasch $Q = Q^r_{max}(R)$ iff $R \hookrightarrow Q$ is a flat epi, and the set D of dense right ideals consists of all right ideals I such $IQ = Q$.*

Proof. See [11], p. 236, Proposition 5.2.

Corollary F. *A commutative Noetherian ring has Kasch $Q_{c\ell} = Q_{max}$.*

Proof. Every dense right ideal of a Noetherian ring R contains a regular element, and this implies that $Q = Q_{max} = Q_{c\ell}$. (See, e.g., [11], p. 237, Example 4.) In particular, this applies to Q itself, so Q has no dense ideals $\neq$ Q, since every regular element is a unit.

The next theorem belongs to Azumaya, Osofsky and Utumi. See [17I] for references, or [3b], Chapter 24.

Theorem G. *For a ring Q the f.a.e.: (PF1) Q is right PF; (PF2) Q is an injective cogenerator in mod-Q; (PF3) Q is right self-injective with finite essential right socle; (PF4) Q is a semilocal right self-injective ring; (PF5) Q is a right Kasch right self-injective ring.*

Remark. (PF4) $\Rightarrow$ Q is a semiperfect ring since idempotents of Q/rad Q lift in a right self-injective ring.

Corollary H. *A commutative ring Q is PF iff Q is a finite product of local self-injective rings with simple essential socle.*

Proof. Q is semiperfect, so the decomposition of Q/rad Q into a finite product of fields lifts to a decomposition of Q into a finite product of local rings which have the stated structure by Theorem G.

2. RANK-ONE GENERATORS ARE PROJECTIVE

If M is a (B, R)-bimodule for a nonempty subset $X \subseteq M$, then

$$r_R X = \{a \in R \mid xa = 0, \ \forall x \in X\}$$

is a right ideal of R, and

$$\ell_B X = \{b \in B \mid bx = 0, \ \forall x \in X\}$$

is a left ideal of B.

If $X \subseteq R$, and $Y \subseteq B$, then

$$\ell_M X = \{m \in M \mid ma = 0, \ \forall a \in X\}$$

is a B-submodule, and

$$r_M Y = \{m \in M \mid bm = 0, \ \forall b \in Y\}$$

is an R-submodule of M.

Let

$$A_r(M, R) = \{r_R X \mid X \subseteq M, \ X \neq \emptyset\}$$

and

$$A_\ell(M, R) = \{\ell_M X \mid X \subseteq R, \ X \neq \emptyset\} .$$

Then the Galois correspondence

$$r_R : A_\ell(M, R) \to A_r(M, R)$$

is 1-1 between the designated (annihilated) B-submodules of M and corresponding (annihilated) right ideals of R, and

$$\ell_M : A_r(M, R) \to A_\ell(M, R)$$

is the inverse mapping.

In case M = R, we set $A_r(R) = A_r(R, R)$, and for any subset $X \subseteq R$, $X \neq \emptyset$, we set

$$X^{\perp} = r_R X .$$

$A_r(R)$ consists of the <u>annihilator right ideals</u> of R or <u>right annulets</u> of R. Similarly, ${}^{\perp}X = \ell_R X$, and $A_\ell(R)$ denotes the set of <u>left annulets</u> of R.

If $X, Y \subseteq Q = Q^r_{max}(R)$, then we set

$$\begin{aligned}(X : Y) &= \{q \in Q \mid qx \in Y, \forall x \in X\} \\ &= \{q \in Q \mid qX \subseteq Y\}.\end{aligned}$$

This symbol is used mainly for M a right R-submodule, in which case (M : R) is a left R-submodule of Q, and then there is a canonical embedding of (M : R)/(M : 0) into the R-dual $M^* = \mathrm{Hom}_R(M, R)$ of M. When this is an isomorphism, we say that Q <u>induces</u> M^*. Then, to each $f \in M^*$ there corresponds $q \in Q$ such that $f(x) = q(x)$, $\forall x \in M$. Furthermore, (M : M) is a subring of Q, and (M : M)/(M : 0) embeds in $\mathrm{End}\, M_R$ canonically. When this embedding is an isomorphism, then we say that Q <u>induces</u> $\mathrm{End}\, M_R$. Then to each $f \in \mathrm{End}\, M_R$, there corresponds $q \in Q$ such that $q(x) = f(x), \forall x \in M$. When Q is injective in mod-R, then each element $f \in M^*$ or $\mathrm{End}\, M_R$ extends to an element $f : Q \to Q$ in mod-R, and then $f(x) = qx$, $\forall x \in M$, where $q = f(1)$. (This by virtue of the canonical isomorphism $Q \approx \mathrm{Hom}_R(Q, Q)$, that is, every R-homomorphism of Q is a Q-homomorphism.) <u>Thus</u>, Q <u>induces both</u> $\mathrm{End}\, M_R$ <u>and</u> M^* <u>when</u> Q <u>is injective.</u>

In the first proposition we are interested in faithful R-bimodules which embed in $Q^r_{max}(R)$. (This proposition is contained in Stenstrom [11], p.149, when M is embedded in R.

1A. Proposition. <u>Let</u> $Q = Q^r_{max}(R)$. <u>If</u> M <u>is any dense right</u> R-<u>sub-module of</u> Q, <u>then</u> $\ell_Q M = 0$, <u>and conversely if either</u> Q <u>is injective or</u> M <u>is an</u> R-<u>bimodule embedded in</u> Q.

Proof. No element of Q except 0 annihilates a dense submodule M, since $\mathrm{Hom}_R(S/M, Q) = 0$ for any R-submodule S containing M. Conversely, if $\ell_Q M = 0$, and if $x, y \in Q$, with $x \neq 0$, then by density of R in Q we may find $r' \in R$ with $yr' \in R$ and $xr' \neq 0$. Then $xr'm \neq 0$ for some $m \in M$, and we can pick $s \in R$ such that $r = r'ms \in R$ and $xr \neq 0$. Since $yr \in M$, we have what we want: Q is rational over M. Next, suppose Q is injective in mod-R. If M is not dense, then $\mathrm{Hom}_R(S/M, Q) \neq 0$, for some submodule S of Q containing M, and hence by injectivity of Q, there is some nonzero $q \in \ell_Q M$ which induces a nonzero R-map $S/M \to Q$.

Corollary. *If R is commutative, then an R-submodule M of Q is dense in Q iff M is faithful.*

The following condition on right ideals M of a right nonsingular ring R,

$$\ell_R M = 0 \Longrightarrow M \text{ is dense } (= \text{essential}) \text{ in } R \text{ (in } Q),$$

is called *right cononsingularity* (e.g., [4]). A theorem of Utumi [26] states that R is right cononsingular iff every nonzero left ideal of Q meets R; equivalently, every closed right ideal of R is a right annulet. Moreover, R is right and left cononsingular iff Q is also the maximal left quotient ring of R. Thus, for non-commutative R, in general $\ell_Q M = 0$ does not imply that M is a dense right R-submodule. However, the proposition shows for R right nonsingular that $\ell_Q M = 0$ implies that M is dense, since then Q is injective.

Corollary. *If R is right nonsingular, then an R-module M embedded in Q is essential iff* $\ell_Q M = 0$.

1B Theorem. *If M is any R-bimodule embedded as a dense R-submodule of* $Q = Q^r_{max}(R)$, *then* $B = \mathrm{End}\, M_R$ *is canonically isomorphic to the subring* $(M : M)$ *of Q, and* $M^* \approx (M : R)$ *canonically. (Thus, Q induces both B and* M^*.)

Proof. Since ${}^{\perp}M = 0$ in Q, then $B = (M : M)$ embeds in Q canonically under $b \mapsto b_s$, where $b_s(m) = bm$, $\forall m \in M$. Hence, in order to prove isomorphism, it suffices to show for any $b \in B$ that any extension of b to an endomorphism β of the injective hull E of Q (in mod-R) is such that $\beta(1) \in Q$, for then $\beta(1)_s = b$. Now, as stated in Section 1, Q has the structure $Q = \text{End}_S E$, where $S = \text{End}\, E_R$, that is, Q is the biendomorphism ring of E, so E is canonically a right Q-module such that $S = \text{End}\, E_Q$. It follows for any x in the right ideal MQ of Q generated by M that $\beta(xq) = \beta(x)q \in MQ$, $\forall q \in Q$. (This follows since $\beta(mq') = \beta(m)q' \in MQ$, $\forall m \in M$, $q' \in Q$.) It therefore suffices to prove the theorem for the case $R = Q$ itself, and to do this, it is necessary and sufficient to prove that $X = tQ+Q$ is a rational extension of Q, for $t = \beta(1)$, for rational closure of Q implies $X = Q$, so then $\beta(1) \in Q$ as required. To do this, let

$$x = tq_1+q_2 \in X, \text{ and } y = tp_1+p_2 \in X \qquad (q_i, p_i \in Q,\ i = 1,2).$$

Let $x \neq 0$, and let $q \in R$ be chosen so that $0 \neq xq \in R$. Then, since $\ell_Q M = 0$, then $xqm \neq 0$ for some $m \in M$, and then $r = qm$ has the property that $xr \neq 0$, and $yr \in M$. (To check the latter, note that $tp_1 = \beta(1)p_1qm = b(p_1qm) \in M$; hence $yr = tp_1qm + p_2qm \in M$.

The proof for $M^* = \text{Hom}_R(M, R) \approx (M : R)$ is similar; that is, $\text{Hom}_R(M,R)$ embeds in $\text{Hom}_Q(MQ,Q)$ canonically. Let b belong to the latter, let β be the extension to S, and let t, x, y, etc., have the same designations as formerly. Then, we still conclude that $xr \neq 0$, and this time $tp_1qm = \beta(1)p_1qm = b(p_1qm) \in R$, so $yr \in R$ as required.

1C Theorem. <u>Let P be any generator over a commutative ring R with commutative endomorphism ring. Then, P is finitely generated projective. In particular, any "rank-1" generator (= a generator embedded in $Q_{max}(R)$) is finitely generated projective.</u>

Proof. By Morita's theorem ([3a], Theorem 7.3), P is finitely generated projective over $B = End_R P$, and hence by Azumaya's theorem (following Proposition 5A), P generates mod-B, hence is finitely generated projective over $End_B P$ which, by another application of Morita's theorem, is R. The second statement then follows from the theorem, since $Q = Q_{max}(R)$ is commutative.

Note, if P is any finitely generated projective over R with $B = End_R P$ commutative, then $B \approx R/ann_R P$ canonically. This follows since P is a finitely generated projective faithful R/A-module for $A = ann_R P$, hence generates mod-R/A, so $R/A = End_B P \supseteq B$ canonically. But, $B \supseteq R/A$ canonically, so equality.

2_{bis} PRE-FP^2F RINGS

Recall that R is right pre-FPF iff every finitely generated faithful right ideal generates mod-R. Pre-FP^2F rings are analogously defined.

1D Corollary. A commutative ring R is pre-FPF (FP^2F) iff every finitely generated (presented) faithful ideal is projective.

Proof. Immediate from the discussion, and Theorem 1C.

1E Theorem. Let R be a commutative ring such that every $b \in Q$ is contained in an R-submodule M_b which generates mod-R. Then $R \hookrightarrow Q$ is a flat epic. In particular, this holds whenever R is FPF. Then, $R \hookrightarrow S$ is a flat epic for any overring S between R and $Q = Q_{max}$.

Proof. This follows from Theorems B, 1B, and 1C, , for in the case at hand, $\phi : R \to S$ is inclusion, and if $b \in S$, then FPF implies that $M_b = bR+R$ generates mod-R, so by Theorem 1B there exist elements $s_i \in M^* \subseteq Q$, $b_i \in M$, such that $\Sigma_{i=1}^n s_i b_i = 1$. Since any $f \in M^*$ maps M, hence R, into R, then f is induced by an element f(1) of R; in particular, $s_i \in R$, and $s_i M \subseteq R$, so $s_i b \in R = \phi(R)$ as required to show that $R \hookrightarrow S$ is a flat epic.

A ring R is <u>right pre-PF</u> provided that every faithful right ideal generates mod-R

A module M is <u>torsionless</u> if it satisfies the equivalent conditions:

TL1. The canonical map $M \to R^{M^*}$ (sending $m \to (\ldots, f(m), \ldots)$) is injective.

TL2. The canonical map $M \to M^{**}$ is an embedding.

TL3. M embeds in a product R^A of copies of R.

1F. Proposition. 1. *A ring R is right pre-PF iff the following two equivalent conditions hold:*

1A. *Every right faithful ideal generates* mod-R.

1B. *Every torsionless faithful right R-module generates* mod-R.

2. *Any left Kasch ring R is right pre-PF.*

3. *A commutative ring R is pre-PF iff every faithful ideal is finitely generated projective.*

Proof. 1. Any right ideal I is torsionless by TL3, so 1B => pre-PF, and trivially pre-PF => 1A. Now assume 1A, and let M be torsionless faithful. If $a \neq 0 \in R$, there exists $x \in M$, $xa \neq 0$, hence $f(xa) \neq 0$ for some $f \in M^*$, so a does not annihilate the trace ideal T(M), that is, T(M) is faithful. Then T(M) generates mod-R, and since T(M) is an epic image of a direct sum of copies of M, then M generates mod-R (so T(M) = R) and 1B holds.

2. A left Kasch ring has no right faithful ideals $\neq$ R, so is right pre-FPF.

3. This is proved similarly to Proposition 1D: Any rank-one generator is finitely generated projective faithful, and conversely.

A ring R has a.c.c. on right annihilators ideals iff the following condition holds:

(acc⊥) For every right ideal I there corresponds a right ideal $I_1 \subseteq I$ such that $\ell_R I = \ell_R I_1$.

We then say that R has acc⊥. (For a proof see [3b], Proposition 20.2A, p. 112.) Let dcc⊥ denote the dcc on right annihilators. Also ⊥acc (resp. ⊥dcc) is the left-right symmetry. Clearly ⊥dcc <=> acc⊥, so for commutative R they are equivalent conditions.

1G. Proposition. 1. A commutative pre-PF ring R has Kasch Q_{max}.
2. Any pre-FPF ring commutative ring R with acc⊥ has Kasch Q_{max}.

Proof. 1. Assume R is pre-PF, and let K be a dense right ideal of Q. Then $M = K \cap R$ is dense hence faithful in mod-R, so by Theorems 1C and 1F, M is a finitely generated projective generator of mod-R, and so, since Q induces M^* by Theorem 1B, then $M^* M = R$ for $M^* \subseteq Q$. Thus, $MQ = Q$, so $K = KQ = Q$, so Q is Kasch by Proposition D(K3).

2. The proof is similar. In this case $\ell_R M = 0 \Rightarrow \ell_R M_1 = 0$ for a finitely generated ideal M_1, which by FPF is projective, and so a generator, and the rest is the same.

3. REDUCED FPF RINGS ARE SEMIHEREDITARY BAER RINGS

The title of this section describes the main result, but for commutative rings only. (The structure of noncommutative nonsingular FPF rings is presently unknown.)

A ring R is reduced if R has no nilpotent elements $\neq 0$.

2A. Proposition. For a commutative ring R, the f. a. e.: (1) reduced; (2) semiprime; (3) nonsingular.

Proof. (1) <=> (2) since semiprime means "no nilpotent ideals $\neq 0$". (2) <=> (3) because in a semiprime ring $x \cap x^{\perp} = 0$ for any ideal x, hence $x \neq 0$ implies $x^{\perp}$ is not essential. Conversely, (3) => (2) since if x is a nilpotent element of R, then $x^{\perp}$ is essential, so (3) => $x = 0$. (If $x^n = 0$, and if y is any element of R, then $xy = 0 \Rightarrow (y) \subseteq x^{\perp}$, and if $xy \neq 0$, then $(xy)^t = 0$ for some least integer $t > 1$, and then $0 \neq y(xy)^{t-2} \subseteq x^{\perp}$. Thus, in either case, $(y) \cap x^{\perp} \neq 0$, so $x^{\perp}$ is essential as stated.)

A ring R is (right) coherent [30] iff R satisfies the two equivalent conditions:

C_1: Every finitely generated (right) ideal is finitely presented.

C_2: If I is a finitely generated (right) ideal, then for all $a \in R$ so is

$(a : I) = \{r \in R \mid ra \in I\}$.

We do not make use of this concept, rather that of pseudocoherence, which is just C_2 for the cases when I is finitely generated, and $a = 0$, that is, ${}^{\perp}I$ is finitely generated for all such I. Actually, we encounter pseudocoherence in a strong form: A ring R is a Baer ring [29] provided that ${}^{\perp}I = Re$, for some idempotent e, for all (one-sided) ideals I. (See Proposition 3B.)

Since any finitely generated projective module is finitely presented (e.g., by Schanuel's lemma [3b], p. 436), then any right semihereditary ring is right coherent. The next result implies a partial converse for FPF reduced rings.

2B. Proposition. Any commutative pseudocoherent reduced pre-FPF ring R is semihereditary.

Proof. If I is any f.g. ideal, then $I^{\perp}$ and $I^{\perp}$ are f.g. By semiprimeness, $I \cap I^{\perp} = 0$, and $I^{\perp} \cap I^{\perp\perp} = 0$, and so $(I+I^{\perp})^{\perp} = 0$; that is, $I+I^{\perp}$ is faithful, hence projective. Since the sum is direct, then I is projective, so R is s.h.

3A. Lemma. In any right FPF ring R, if I and K are right ideals such that $I \cap K = 0$, then $R = {}^{\perp}IR + {}^{\perp}KR$. Thus, if A and B are ideals such that $A \cap B = 0$, then $R = {}^{\perp}A + {}^{\perp}B$. If $A = {}^{\perp}B$ is an annihilator ideal and if $A \cap {}^{\perp}A = 0$, then $R = {}^{\perp}A \oplus A$, hence A is generated by a central idempotent.

Proof. The module $M = R/I \oplus R/K$ is faithful, hence is a generator of mod-R, so its trace ideal $= {}^{\perp}IR + {}^{\perp}KR = R$. (Hint: ${}^{\perp}I$ is the dual module of R/I, and hence ${}^{\perp}IR$ is the trace ideal of R/I.) Thus, if A is an ideal, ${}^{\perp}AR = {}^{\perp}A$ is an ideal, so $R = {}^{\perp}A + {}^{\perp}B$ as asserted. Suppose $A = {}^{\perp}B$, and write $1 = e + f$, with $e \in A$, $f \in {}^{\perp}A$. Then, $e = e^2$, and also $f = f^2$, and, moreover, $a = ea$, $\forall a \in A$, since $fa = 0$. However, since $AB \subseteq A \cap B = 0$, then $B = {}^{\perp}A \subseteq A^{\perp}$, so that $a = ae$, and $af = 0$, proving that $A = eR$ is generated by a central idempotent e.

3B. Proposition. In a semiprime right FPF ring, every annihilator ideal is generated by a central idempotent, and any reduced commutative FPF ring is a semihereditary Baer ring.

Proof. Let $A = {}^{\perp}B$ be an annihilator ideal. Since R is semiprime, then $A \cap B = 0$, so A is generated by a central idempotent by the lemma, and hence, then so is $A^{\perp}$, that is, every annihilator ideal is generated by a central idempotent.

Since a commutative ring R is reduced iff semiprime (and iff nonsingular) then a commutative reduced FPF ring is Baer, hence s.h. by Proposition 2.

3C. Corollary. A directly indecomposable semiprime right FPF ring R is prime.

Proof. If R is semiprime, then $A \cap {}^{\perp}A = 0$ for any two ideals, so Lemma 3A and the fact that R is indecomposable implies that either $A = 0$ or ${}^{\perp}A = 0$, that is, that R is prime.

3D. Example. For any finite group $G \neq 1$ the integral group ring is never FPF, since it is indecomposable but not prime.

A ring R is right Rickart if $x^{\perp}$ is generated by an idempotent for each $x \in R$. Thus, any Baer ring, and any semihereditary ring is right Rickart.

3E. Lemma. A commutative ring R has a von Neumann regular quotient ring $Q = Q_{c\ell}(R)$ iff for each element $a \in R$ there corresponds $x, y \in R$ with y regular such that $axa = ay$ $(= a^2x)$. Any commutative Rickart ring R has regular $Q_{c\ell}$.

Proof. A ring Q is von Neumann regular iff for every $a \in Q$ there corresponds $x \in Q$ such that $axa = a$, and this is equivalent to the requirement that $aQ = eQ$ for some idempotent $e \in A$. (For $axa = a$, use $e = ax$.) Now a typical element of Q has the form au^{-1}, with $a, u \in R$, and u regular, and $au^{-1}Q = aQ$, so that Q is regular iff the condition $aQ = eQ$ holds for all $a \in R$. When this is so, then $e = axy^{-1}$ with $x, y \in R$, and then $ay = axa$, etc. If R is Rickart, and if $f = f^2$ generates $a^{\perp}$, then $y = a+f$ is regular, and $a^2 = ay$, so $axa = ay$ with $x = 1$.

4. Theorem. For a semihereditary commutative ring, the f. a. e.:

(1) R is FPF.

(2) $\overline{R} = Q_{max}$ is a flat epic.

(3) $Q_{c\ell}$ is injective.

(4) $Q_{c\ell} = Q_{max}$.

Proof. (1) ⟹ (4). By Theorem C, $Q = Q_{c\ell}$ is FPF, and by Lemma 3E, Q is regular. Now, Q_{max} is flat epic over Q by Theorem 1E, and, as is well known, a von Neumann regular ring Q is an "epic final" ring in the sense that a ring embedding $R \hookrightarrow T$ is epic iff it is an isomorphism. (Compare [11], p. 225, Proposition 1.4.) Thus, $Q = Q_{max}$. This can also be seen by using Theorem 1C to see that every finitely generated Q-module M between Q and Q_{max} is projective, and then use the unimodular row lemma[4] to conclude that Q is a direct summand, hence = M, so $Q = Q_{max}$.

(4) ⟹ (3) since Q_{max} is injective.

(3) ⟹ (2) since $Q_{c\ell}$ injective ⟹ $Q_{c\ell} = Q_{max}$, and $Q_{c\ell}$ is flat epic over R.

(2) ⟹ (1). Let M be any f. g. faithful module. Then, since R is a n. s. ring, $\overline{M} = M/\text{sing}\, M$ is a nonsingular module, and hence, by the theorem of Cateforis [2] (cited in Section 1), $\overline{M}$ is projective, so $M = A \oplus K$, where $K = \text{sing}\, M$, and $A \approx \overline{M}$. Now $\text{ann}_R M = \text{ann}_R A \cap \text{ann}_R K = 0$. Since K is f. g. and singular, then $\text{ann}_R K$ is an essential ideal; hence $\text{ann}_R A = 0$. Therefore, $\overline{M} \approx A$ is a faithful f. g. projective module, which by the theorem of Azumaya [1] stated in the proof of Theorem 1C implies that $\overline{M}$, whence M, generates mod-R. This proves R is FPF.

Remarks. 1. Not every s. h. ring can be FPF in view of Theorem 4,

---

[4] Consult, e. g., [3a], p. 180.

e.g., a von Neumann regular ring which is not injective, or, equivalently any Boolean ring $\neq$ its completion.

2. Not every semihereditary Baer ring is FPF. This follows from Utumi's Example 3 of [33] and Theorem 4, since there exist commutative regular Baer rings which are not self-injective.

4. LOCALLY (C)FP^{2}F RINGS ARE (C)FP^{2}F RINGS

The free rank of a module M, denoted frk M, is the largest integer n such that at every maximal ideal P of R the local module M_P contains a free direct summand of rank n. If no such exists: in case some M_P has no free direct summand, put frk M = 0; or if for every P, M_P has a free direct summand of every rank, put frk M $= \infty$.

The gist of the next well-known proposition is that a finitely presented module is a generator iff it is locally a generator. (I am indebted to W. Vasconcelos for this reference.)

5A. Proposition. Over a commutative ring R, a finitely presented module M is a generator iff frk M ≥ 1.

Proof. Let P be a prime ideal. Since R_P is local, M_P generates mod-R_P iff M_P has a free direct summand. Thus, since this happens for all prime ideals P iff it happens for the maximal ideals, this is equivalent to stating that frk M ≥ 1.

For any f.p. R-module M, and flat overring T of R we have a functorial isomorphism

$$\mathrm{Hom}_R(M,X) \otimes T \approx \mathrm{Hom}_S(M \otimes_R T, X \otimes_R T)$$

for any R-module X. Thus, the "dual of M" commutes with localization. This is the statement above for X = R, and $T = R_S$ for any multiplicatively closed subset S of R,

$$(\mathrm{Hom}_R(M,R))_S \approx \mathrm{Hom}_{R_S}(M_S, R_S).$$
$$(M^*)_S = (M_S)^*$$

Taking S to be the complement of a prime ideal P, we have at any local ring R_P of R that "the dual of the local module M_P is the local module of the dual of P." Similarly for the trace ideal T(M) of M, $T(M)_S = T(M_S)$. Thus for an f.p.

module M, $T(M_P) = R_P$, $\forall P$, iff $T(M) = R$, so $M_P = M \otimes R_P$ generates mod-R_P, $\forall P$, iff M generates mod-R.

Corollary (Azumaya's Theorem). Over a commutative ring R any faithful finitely generated projective module P is a generator.

Proof. "Faithful", "finitely generated" and "projective" are local properties, finitely generated projective are finitely presented, and over local rings, nonzero projectives are free. Since Γ rk $P \geq 1$, then P is a generator.

For what follows we need several definitions.

A module M is uniserial if the lattice of submodules is linearly ordered (= a chain). A ring R is a right valuation ring (right VR) if the right R module R is uniserial. A commutative ring R is a VR iff finitely presented modules are direct sums of cyclics, and locally a VR in the sense that R_M is a VR for every maximal ideal M iff every finitely presented module is a direct summand of a direct sum of cyclics ([31, 32]).

A valuation ring R is said to be almost maximal [31] provided that any system $X \equiv X_{\alpha}$ (mod I_{α}) of congruences, for ideals $\{I_{\alpha}\}$ and elements $\{X_{\alpha}\}$, is solvable iff finitely solvable and $\bigcap I_{\alpha} \neq 0$. (Maximal VR is AMVR without the restriction on the intersection.)

A ring R is FP^2F if every finitely presented faithful module M is a generator of mod-R. A ring R is CFP^2F if every factor ring is FP^2F, and similarly for CFPF.

5B. Theorem. Let R be a commutative local ring.

1. The following conditions are equivalent:

   (a) R is a VR.

   (b) R is CFP^2F.

(c) Every finitely presented module is a direct sum of cyclic modules.

2. The following conditions are equivalent:

(a) R is an almost maximal valuation ring (AMVR).

(b) Every finitely generated module is a direct sum of cyclic modules.

(c) The minimal injective cogenerator E(R/rad R) is uniserial.

(d) R is CFPF.

Proof. In both 1 and 2, the statements not involving FP^2F, CFP^2F, or CFPF are theorems of Krull, Kaplansky, and Warfield (in 1), and Kaplansky and Gill (in 2). (This is discussed, and references given, in [3b], Chapters 20 and 25.) The other statements are trivial consequences, e.g., if every finitely generated (presented) module M is a direct sum of cyclics, then M faithful implies in a VR that at least one of the cyclic $\approx$ R; hence M is then a generator. Since every factor ring has the same property, then R is CFPF (resp. CFP^2F). Thus, 1(a) => 1(b), and 2(b) => 2(d). Conversely, if R is FP^2F local, then R is a uniform ring by the following lemma, and if R is CFP^2F, then every factor ring is uniform. This implies R is uniserial, since for ideals A and B, in the residue ring $R/(A \cap B)$, A or B maps onto 0. This completes the proof of 1(a) <=> 1(b).

Finally, assuming 2(d), if M is any finitely generated module, and if $A = \text{ann}_R M$, then M is faithful over R/A, hence generates mod-R/A, which over a local ring implies by the Krull-Schmidt or Exchange Lemma that $M = R/A \oplus X$, for some finitely generated submodule X. By induction on the vector space dimension of M/MJ over R/J, where J = rad R, we conclude that M is a direct sum of cyclic modules, that is, 2(b) <=> 2(d).

5C. Lemma. Any (right) FP^2F local ring R is (right) uniform (= R has indecomposable injective hull.)

Proof. The conclusion is equivalent to the implication $A \cap B = 0 \Rightarrow A = 0$ or $B = 0$ for f.g. right ideals A, B. But this follows as in the proof of Lemma 3A, since the trace ideal of $R/A \oplus R/B$ is $R = {}^{\perp}AR + {}^{\perp}BR$, which in a local ring implies that ${}^{\perp}A$ on ${}^{\perp}B$ contains a unit, whence ${}^{\perp}A = R$ and $A = 0$, or $B = 0$.

5C'. Generalization. <u>In a semiperfect right FP^2F ring R, every principal indecomposable right ideal is uniform.</u>

Proof. Same proof as Theorem 1 of [17I].

Let R^* denote the set of regular elements of R.

5D. Proposition. <u>Let R be a commutative ring. Then:</u>

1) <u>If $R \hookrightarrow R_P$, that is, if $R \setminus P \subseteq R^*$, where P is a prime ideal,</u> R FPF (FP^2F) <u>implies that R_P is FPF (FP^2F).</u>
2) <u>If R is CFPF (CFP^2F), then R is locally CFPF (CFP^2F).</u>
3) <u>If R is locally (C)FP^2F, then R is (C)FP^2F.</u>
4) <u>R is CFP^2F iff R is locally CFP^2F.</u>
5) <u>A domain R is FP^2F iff R is locally FP^2F.</u>

Proof. 1) follows from Theorem C (§1) for FPF.

If R is FP^2F, and if M is f.p. (faithful) over R_P for a prime ideal P, then $M = N \otimes_R R_P$ for an f.p. (faithful) R-module N. (In effect, if $R_P^n \xrightarrow{f} R_P^m \to M \to 0$ is exact over R_P, then the mapping f is given by an $m \times n$ matrix $(a_{ij}b^{-1})$, with $a_{ij} \in R$, and $b \in R$, and then there is an f.p. R-module N fitting into an exact sequence $R^n \xrightarrow{(a_{ij})} R^m \to N \to 0$ and clearly $M = N \otimes_R R_P$.) Since $R \hookrightarrow R_P$, then N is faithful if M is, and therefore N generates mod-R by FP^2F, and hence, then M generates mod-R_P.

2) essentially follows from 1), since if K is the kernel of the canonical map $R \to R_P$, for any prime ideal P, then $P \supseteq K$, and $K_P = K \otimes_R R_P = 0$, so $(R/K)_{P/K} = R_P/K_P = R_P$ canonically. Now $R/K \approx (R/K)_{P/K}$, and R/K is FPF (FP^2F), and

therefore by 1) so is $R_P \approx (R/K)_{P/K}$. Moreover, if A is any ideal of R_P, then $A = I_P$ for any ideal I of R, and then $R_P/A \approx (R/I)_P$ is FPF (FP^2F) since R/I is CFPF (CFP^2F).

3) follows from Proposition 5A for FP^2F. Specifically, if M is any f.p. faithful R-module, then M_P is f.p. faithful over R_P for any prime ideal P, and hence generates mod-R_P by locally FP^2F hypothesis, so Proposition 5A implies that M generates mod-R.

To complete the proof for a locally CFP^2F ring R, let I be any ideal, and let P/I be any prime ideal of R/I. Since R_P/I_P is FP^2F, and $(R/I)_{P/I} \approx R_P/I_P$ canonically, then R/I is locally FP^2F; hence R/I is FP^2F. Therefore R is CFP^2F.

4) is the resultant of 2) and 3), and 5) is that of 1) and 4).

5E. Corollary. _A ring R is CFP^2F iff R is locally a valuation ring._

Proof. Apply 5D and 5B.

5F. Corollary. _If R is a locally (C)FPF, then R is (C)FP^2F._

Proof. For then R is locally (C)FP^2F, so 5D applies.

We note that an FPF local ring need not be a VR. (See Example 9E.) Therefore, by 5B, FPF does not imply CFP^2F. Furthermore, a locally CFPF ring need not be FPF inasmuch as a commutative von Neumann regular ring R is locally a field, but by Theorem 4 is FPF iff self-injective[5]. Finally, an FPF ring need not be locally FPF! (Example 16E).

A ring R is a <u>flat ideal</u> ring if every ideal is a flat R-module. The following is known, and easy to prove.

5G. Proposition. <u>For a commutative ring, the f. a. e.</u>:

(a) R <u>is a flat ideal ring.</u>

(b) R <u>is locally a valuation domain (VD).</u>

5H. Corollary. <u>Any commutative flat ideal ring</u> R (<u>e. g.</u>, <u>any semi-hereditary ring</u>) <u>is</u> CFP^2F.

Proof. Apply 5E.

5I. Examples.

(a) Prüfer rings are the domains which are locally VD's; equivalently s. h.

(b) Dedekind rings are the Noetherian Prüfer rings.

(c) Any commutative quasi-Frobenius (QF) ring R is the finite product of local Artinian VR's. Since any factor ring of a commutative QF ring has the same property, these rings are CFPF.

(d) Every proper factor ring of a Dedekind domain is QF, and hence Dedekind domains are CFPF.

(e) A domain R is FP^2F iff f. p. ideals $\neq 0$ are projective, and iff in every R_P f. p. ideals $\neq 0$ are principal. (See Corollary 1D and Proposition 5D(5).)

---

[5] This fact was communicated to me by R. Wiegand. Note in the example, R_P is actually QF for any P.

5. LOCAL FPF RINGS ARE QUOTIENT-INJECTIVE

In this section local FPF rings are characterized by the condition that $Q_{c\ell}$ is injective and the set P of zero divisors of R is a waist such that R/P is a VR. As a corollary we obtain a characterization of when a local ring R is self-injective, viz. $R = Q_{c\ell}$ and FPF. Thus, if rad R = P, and R is FPF, then R is injective; in other words, $Q_{c\ell}$ is injective iff FPF. These two conditions are ideal-theoretic in the sense that the first states that the only principal right ideal $(x) \approx R$ is R, and the second states that the trace ideal of every finitely generated faithful module is the unit ideal.

In general a flat module F, although a direct limit of projective modules, need not be a sum of projective submodules. If it is, we say that F is projectively generated. If, furthermore, F is a sum of free submodules, we say that F is freely generated.

6A. Proposition. If R is commutative, and $\overline{R} = Q_{max}(R)$ is freely generated, then $Q_{max} = Q_{c\ell}$.

Proof. Let M be any free submodule of Q. By Theorem 1B, End M_R is induced by $\overline{R}$, and since $\overline{R}$ is commutative, so is End M_R; hence M must be free on one generator, say $M = xR \approx R$ canonically. Now $1 \in M$, so $1 = xa$ for some $a \in R$, and then a is not a zero divisor of R. This implies that $a^{-1} \in Q_{c\ell}$, and hence, $x = a^{-1} \in Q_{c\ell}$, so $M \subseteq Q_{c\ell}$. This implies that $\overline{R} \subseteq Q_{c\ell}$, and completes the proof.

6B. Proposition. If R is a commutative ring of Goldie dimension $n < \infty$ and freely generated injective hull E, then $Q_{max} = Q_{c\ell}$ is injective (= E). (Hence n = 1.)

Proof. If M is any free submodule of rank c, and if $c > 1$, then E contains a direct summand $\approx E^2$, which is impossible since Goldie dimn $< \infty \Rightarrow E$ satisfies the Krull-Schmidt theorem because $E = F_1 \oplus \dots \oplus F_n$ where F_i is indecomposable with local endomorphism ring, $1 = i, \dots, n$. (See, e.g., [3b], 18.15.) Thus, $c = 1$, so that $M \approx R$, and then the proof of Proposition 6A shows that $M \subseteq Q_{c\ell}$, and that $E = Q_{c\ell}$. Since $E \supseteq Q_{max}$, this proves what we want.

7. Proposition. <u>If R is a commutative FPF local ring, then</u> $Q_{c\ell}(R)$ <u>is injective.</u> (<u>Cf. Theorem</u> 9B.)

Proof. The proof is similar to the theorem of Faith-Zaks [5] (which is the special case when R is a VR). If E is the injective hull of R and M is a f.g. submodule $\supseteq R$, then M generates mod-R, so $M \approx R \oplus X$ for a submodule X. But, <u>any</u> local FPF ring R is uniform by Corollary 5C. Hence $X = 0$ and $M \approx R$ holds. Then, since R has Goldie dimension 1, Proposition 6B implies $E = Q_{c\ell}(R)$ is injective.

8. Example. If R is a local Artinian ring not injective (equivalently not a principal ideal ring = not a VR), then clearly $R = Q_{c\ell}(R)$. But R Noetherian $\Rightarrow Q_{c\ell}(R) = Q_{max}$. However, this ring is not injective, that is $Q_{max} \neq E$.

A submodule W of a module M is a <u>waist</u> if either $S \supseteq W$ or $S \subseteq W$ for every submodule S. Clearly, M and O are waists of any module. If a finitely generated module M has a unique maximal submodule M', then M' is a waist since every submodule $\neq M$ embeds in a maximal submodule. Thus, rad R is a waist of a local ring R. Similarly, an essential simple submodule is a waist. (These are trivial examples of waists, however.) Clearly, every submodule is a waist iff M is <u>uniserial</u> in the sense that the lattice of submodules is linearly ordered. In the next theorem, local FPF rings are characterized via three conditions, including the condition that the set P of zero divisors (z. d. s.) is a waist. (In a uniform ring, e.g., an FP^2F local ring (5C), the set P of zero divisors is an ideal since $x^{\perp} \neq 0$ and $y^{\perp} \neq 0 \Rightarrow x^{\perp} \cap y^{\perp} \neq 0$, so $(x+y)^{\perp} \neq 0$.)

In order to characterize the local FPF rings it is convenient to utilize the next lemma.

9A. Lemma. Over a local ring R, for a module M and submodule W the following conditions are equivalent.

(1) For all $x, y \in M \setminus W$, either $xR \supseteq yR$, or $yR \subsetneq xR$.

(2) M/W is uniserial and W is a waist.

Proof. (1) => (2). If $x \in M$, $p \in W$, then $x \notin W \Rightarrow x+p \notin W$. Now $xR \subset (x+p)R$ implies $x = (x+p)a$ for some $a \in R$, and $x(1-a) = pa$. Then $1-a$ is not a unit since $x \notin W$, so a must be a unit, and hence only the other case $(x+p)R \subseteq xR$ can hold. Then, $x+p \in xR$, whence $p \in xR$, so $xR \supseteq W$.

Conversely, assume (2). If $x, y \in M \setminus W$, then $\bar{x}R \supseteq \bar{y}R$ or $\bar{y}R \supseteq \bar{x}R$ = in the factor module $\bar{M} = M/W$. In the first case, $y = xa+p$, for some $a \in R$, and $p \in W$. But W is a waist, so $xR \supset W$, so $y = xa + p \in xR$. Similarly in the other case, proving (1).

9B. Theorem. For a local ring R, the f.a.e.:

(1) R is FPF.

(2) Q_{cl} is injective and every f.g. module M between Q_{cl} and R is cyclic.

(3) Q_{cl} is injective and the set P of zero divisors of R is a waist such that R/P is a VR.

Proof. First assume (1). Then R is uniform by Corollary 5C. Moreover, any f.g. submodule M between R and $Q = Q_{cl}$ is cyclic because M generates mod-R, hence $M \approx R \oplus X$, but M is uniform, along with R, hence $X = 0$, so $M \approx R$ is cyclic.

Next, assume (2). If $x \in Q$, $x \notin R$, then $xR+R$ is cyclic, say $xR+R = yR$. Write $x = ya$, and $y = xc+b$ for $a, b, c \in R$. Then $y = xac+b$, so $y(1-ac) = b$. Since $x \notin R$, then $y \notin R$, so $1-ac$ is not a unit of R, that is, $ac \notin \operatorname{rad} R$, so ac is a unit of R. This implies that a is a unit of R, so

(1) $$xR = yaR = yR = xR+R.$$

We next prove, for any two elements, $u, v \in Q$,

(2) $$u \notin R \Rightarrow uR \supseteq vR \quad \text{or} \quad vR \subseteq uR.$$

To see this, we have $uR = R+uR$ by (1), so u is not a zero divisor in Q. Then $u^{-1} \in Q$, and by (1) we have

(3) $$u^{-1}v \notin R \Rightarrow u^{-1}vR = R+u^{-1}vR \Rightarrow vR = uR+vR \subseteq uR$$

proving (2) in this case. In the other case

(4) $$u^{-1}v = b \in R \Rightarrow vR = ubR \subseteq uR,$$

so (2) is complete.

By the remark preceding the statement of the theorem, P is an ideal, and if $x, y \in R \setminus P$, then, by (2),

(5) $$x^{-1} \notin R \Rightarrow xR \supseteq yR \quad \text{or} \quad yR \subseteq xR$$

and

(6) $$x^{-1} \in R \Rightarrow xR = R \supseteq yR,$$

so R/P is a VR and P is a waist by Lemma 9A. This proves (3).

Finally, assume (3). Let $Q = Q_{c\ell}(R)$. Clearly, Q is the local ring R_P. The proof follows the proof in [5] of the corresponding result in case R is a VR, and we include the short proof for completeness. If M is f.g. faithful over R, then the torsion submodule $t(M)$ of all elements annihilated by regular elements cannot be all of M, and $\overline{M} = M/t(M)$ inherits the hypothesis on M, namely, f.g. faithful. But if $\overline{M}$ generates mod-R, so will M; hence we may assume M is torsion free or equivalently, $M \approx M \otimes_R R \hookrightarrow M \otimes_R Q = M_2$ canonically. Since any self-injective local ring is FPF, then M_2 generates mod-Q, so there is an epic map $f : M_2 \to Q$. If $m_1, \ldots, m_n$ generate M, then $f(m_i) = r_i t^{-1}$ for some $r_i \in R$, $t \in R$, with $t \notin P$, $i = 1, \ldots, n$. Since some $r_i \notin P$, renumber $m_1, \ldots, m_n$ so that $(r_n) \supseteq (r_i)$ for any $i \leq n$. (This can be done in view of Lemma 9A.) Then, clearly, if t_d is the multiplication of M by t, and $\overline{f} = f|M$, we have that the image of $t_d\overline{f}$ is $\Sigma_{i=1}^{n} tr_i R = (r_n) \approx R$. Thus, M maps epically onto a generator R, so M is a generator. Thus (3) $\Rightarrow$ (1), proving the theorem.

9C. Corollary. _A commutative uniform FPF ring R has injective local $Q_{c\ell}$._

Proof. As stated in 9A, the set P of zero divisors is an ideal, and hence $Q_{c\ell} = R_P$ is a local ring which by Theorem 1E is FPF, and which by Theorem 9A is injective.

9D. Proposition. _For a commutative ring R the f.a.e.:_

(1) _R is a self-injective directly indecomposable ring._

(2) _R is a self-injective uniform ring._

(3) _R is a uniform FPF quotient ring._

When this is so, then R is a local ring.

Proof. An injective module is indecomposable iff uniform iff the endomorphism ring is local. Thus, (1) $\Leftrightarrow$ (2), and then $R = \mathrm{End}_R R$ is local. Since every regular element x in a self-injective ring R is a unit, then, via Footnote

4, (1) ⟹ (3), and the converse is Corollary 9C.

9D′. Generalization of 9D. _If R is any semiperfect (commutative or noncommutative) right FPF ring such that for every primitive idempotent e the radical of eRe consists of zero divisors, then R is right self-injective._

Proof. See the proof of Theorem 1 of [171], where the result is stated for the case rad R is nil.

9E. Example. Let A any Noetherian local domain, let E be the injective hull of the unique simple module, and let $B = \mathrm{End}_A E$. Then, by [18], there is a self-injective ring R consisting of all 2×2 matrices $(b,x) = \begin{pmatrix} b & x \\ o & b \end{pmatrix}$ with $b \in B$, $x \in E$, with ordinary matrix multiplication and addition. Then R is FPF by 9D, but R is a VR iff A is a principal ideal domain (PID), that is, iff A is a VR.

6. THE SPLITTING THEOREM AND THE THEOREM OF ENDO

The next theorem, needed in our proof of Endo's theorem [25], is for a general FPF ring R.

10. Splitting Theorem. *Any right FPF ring R splits into a product $R_1 \times R_2$ where R_1 is a ring without nil (resp. nilpotent) ideals, and if N (resp. N_0) is the nil radical (resp. the sum of the nilpotent ideals) of R, then N (resp. N_0) meets every nonzero ideal of R_2. If R_1 is commutative, then R_1 is s.h., and $N(R_2) \subseteq \operatorname{sing} R_2$.*

Proof. Let B be an ideal maximal w.r.t. $B \cap N = 0$. Let $A = {}^{\perp}B$. Since $(A \cap B)^2 = 0$, then $A \cap B \subseteq N \cap B = 0$; and hence by Lemma 3D, $R = R_1 \times R_2$, where $R_1 = {}^{\perp}A$, and $R_2 = A$.

If I is any ideal of R_2 and if $I \neq 0$, then $(B+I) \cap N \neq 0$, so if $x = b+y$, with $x \in N$, $b \in B$, $y \in I$, then $x \neq 0 \Rightarrow b \neq 0$ and $y \neq 0$, and, moreover, $b = x-y \in B \cap A$, so $b = 0$, and $x = y \neq 0$ is an element of $I \cap N$. This proves that every ideal $I \neq 0$ meets N.

If R_1 is commutative, then R_1 is reduced FPF along with R, hence is s.h. by Proposition 2. Since the annihilator ${}^{\perp}I$ is an essential right ideal whenever I is nilpotent, then clearly when R is commutative, every nilpotent element, whence N, is contained in sing R.

The proof for N also suffices for the proof for N_0.

The next theorem is the theorem of Endo mentioned in the introduction. Our proof is more elementary, avoids valuation theory, but depends on a theorem of Robson which was not available to Endo. The theorem of Robson [14] states: A Noetherian ring Q splits into a product of a semiprime ring and an Artinian iff it is true that $cN = Nc = N = N(Q)$ for any $c \in R$ which is regular modulo N. Our proof also requires the Krull Intersection Theorem.

11. Theorem. (Endo [25]). *Any Noetherian FPF commutative ring is*

quotient-injective, and $R = R_1 \times R_2$, where R_1 is a finite product of Dedekind domains, and R_2 is Quasi-Frobenius (QF). Thus, $Q_{c\ell}(R)$ is QF.

Proof. By Theorem 10, $R = R_1 \times R_2$, where R_1 is semihereditary and $N(R_2)$ essential in R_2. Any Noetherian semihereditary ring is a finite product of Dedekind domains, so it remains to prove $Q_{c\ell}(R_2)$ is QF. By Theorem C, any FPF ring R has FPF $Q_{c\ell}$. (In fact, any quotient ring B of R is FPF provided that $R \hookrightarrow B$ is an embedding.)

The fact that $Q_2 = Q_{c\ell}(R_2)$ is FPF can be used to prove that Q_2 is Artinian using Robson's theorem. Actually, any $c \in R$ such that $c^{\perp} \subseteq N = N(Q_2)$ must be regular, hence $c^{-1} \in Q_2$, and therefore $cN = N$, which is Robson's condition for Q_2 to be Artinian. The reason c is regular is that there is an n such that $(c^n)^{\perp} \cap (c^n) = 0$ and then $R = (c^n)^{\perp\perp} + (c^n)^{\perp}$ by Lemma 3A. But $(c^n) \subseteq N \Rightarrow R = (c^n)^{\perp\perp}$, whence $(c^n)^{\perp} = 0$, so $c^{\perp} = 0$.

Now that Q_2 is Artinian FPF, we could apply Tachikawa's theorem [15] to conclude that Q_2 is QF. On the other hand, we can deduce this from the special case of Theorem 9B since an Artinian commutative FPF ring splits into a finite product of local Artinian FPF rings, each of which must be injective by 9B, since they each are their own $Q_{c\ell}$. Since an injective Artinian ring is QF, this means Q_2 is QF.

To complete the proof, we must show that $R_2 = Q_2$, and to do this, it suffices to prove that R_2 is Artinian. By Theorem 10 we may assume that $R = R_2$ has essential nil radical $N = N(R)$ and that N is nilpotent since R is Noetherian. Since $Q = Q_{c\ell}(R)$ is QF, Q is a finite product $\prod_{i=1}^{n} Q_i$ of local QF rings. Let $B = \prod_{i=1}^{n} R_i$, a subring of Q containing R, where $R_i = e_i R$ and e_i is the identity element of Q_i, $i = 1, \ldots, n$. By the parenthetical remark in the first paragraph of the proof, B is FPF, and hence each R_i is FPF with local $Q_{c\ell} = Q_i$, $i = 1, \ldots, n$.

Now R is Artinian iff $R = Q$, hence iff $R_i = Q_i$, $i = 1, \ldots, n$, so it suffices to prove the theorem for the case Q is a local QF ring. Since $P = \operatorname{rad} Q$ is nilpotent, clearly, $P = NQ$, and N is the set of zero-divisors of R. Thus, if M is a maximal ideal of R containing $J = \operatorname{rad} R$, the local ring R_M embeds in Q, and hence is FPF.

This reduces to the case R is local FPF, since R_M Artinian implies that $R_M = Q$, hence that $M = N$ is the unique maximal ideal. By Theorem 9B, N is a waist of R, and R/N is a VR; hence some $x \in R$ generates J modulo N (since R is a Noetherian VR modulo N). But N is a waist, so $J = xR$. Then, assuming $J \neq N$, we have $x^n \notin N$ for every n; hence N is contained in the intersection of all of the powers J^n, which by the Krull intersection theorem is zero. But N is essential in R, a contradiction which proves the assertion $J = \operatorname{rad} R = N$.

12A. Corollary. _A commutative Noetherian FPF ring R with essential nil radical is QF._

12B. Corollary. _Any directly indecomposable Noetherian commutative FPF ring R is either a Dedekind domain, or an Artinian VR. In either case, R is CFPF._

Proof. Apply Theorem 11, and use the fact that any proper factor ring of a Dedekind domain is QF. (Also, any commutative QF ring has the same property!)

12C. Corollary. _Any Noetherian commutative CFPF ring R is FSI._

The next theorem generalizes 12B.

12B. Generalization. _If R is a 2-sided Noetherian semiperfect FPF ring, then $R = R_1 \times R_2$, where R_1 is a finite product of bounded Dedekind prime rings, and R_2 is QF._

Proof. See [16], Theorem 13.

13. Proposition. _A Noetherian locally FPF ring R is CFPF._

Proof. A Noetherian local FPF ring R is a valuation ring by Corollary 12A, hence is CFP^2F (by Corollary 5B). Since R is Noetherian, then R is $CFP^2F = CFPF$ by Proposition 5D.

14. Remark. Any Noetherian CFPF ring R splits, $R = R_1 \times R_2$, where R_1 is a finite product of bounded Dedekind rings, and R_2 is QF. (See [4].) Thus, any Noetherian locally FPF ring has this structure by the proposition.

15. Problem. Let R be FPF (or CFPF).

15A. Is $Q_{c\ell} = Q_{max}$?

15B. Is $Q_{c\ell}$ injective?

15C. Is Q_{max} injective?

15D. Does 12B hold assuming only that R is 2-sided Noetherian FPF? (I.e., without assuming semiperfect?) (Cf. 14.)

15E. When does a ring R decompose into a product $R_1 \times R_2$ with R_1 semiprime and R_2 injective? This of course holds when R is FPF and Noetherian commutative (Theorem 11) or Noetherian semiperfect (Theorem 12B). The truth of our conjecture(s) that any FPF ring is quotient-injective (resp. $Q_{max}(R)$ injective) would yield the characterization that $R_2 = Q_{c\ell}(R_2)$ (resp. $Q_{max}(R)$) when R is FPF.

16. Example. (An FPF ring need not be locally FPF.) Let A be any Noetherian local domain, let E denote the injective hull of the unique simple module A/J, where J = rad A. Then, by a theorem of Matlis [22], $B = \text{End}_A E$ is the completion (in the J-adic topology) of A, E is the injective hull of B/rad B, and and $E = \text{End}_B E$. Now, as stated in 9E, the semidirect product ring R = (B, E) is self-injective; hence, by Proposition 9D, R is FPF. Now suppose R is locally FPF, that is, R_P is FPF for every prime ideal P. Then, $B \approx R/(O, E)$ would be locally FPF by the proof of 5D; hence, by Proposition 13, B would be CFPF. But, CFPF in a local ring B is equivalent to B being an almost maximal valuation ring which, since B is a Noetherian local domain, is equivalent to B being a PID. Thus: <u>A general FPF local ring R cannot be locally FPF.</u> A domain A is FPF iff A is Prüfer; hence, any FPF domain is locally FPF.

17. Remark 1. Semiperfect right CFPF rings have been determined in [4] as finite products of full matrix rings over rings which are right VR's right duo and right σ-cyclic.

Remark 2. By the example of [18], we know that local self-injective rings need not be PF, hence an FPF ring need not have PF $Q_{c\ell}$ or Q_{max}. If, however, $Q = Q_{c\ell}$ or Q_{max} is Kasch, then Q will be injective iff PF by Theorem F of §1 (and then, by Corollary H, Q is a finite product of local PF rings). In particular, then Q whence R has finite Goldie dimension. In regard to Question 15, then we ask: Is an FPF Kasch ring injective, that is, PF?

7. SIGMA CYCLIC RINGS AND VAMOS' THEOREM

A ring R is right σ-cyclic (= FGC) if every f. g. right R-module can be written as a direct sum of cyclic modules. These rings include all principal ideal rings, almost maximal valuation rings (AMVR's) and their finite products. Recently σ-cyclic commutative rings have been determined as finite products of rings having the four properties:

(FGC1) R has a unique minimal prime P.

(FGC2) R_M is an AMVR for each maximal ideal M.

(FGC3) R/P is an h-local Bezout domain.

(FGC4) The ideals of P form a chain.

Bezout means that every f. g. ideal of R is principal. And h-local means every nonzero element is contained in just finitely many maximal ideals.

The classification FGC1-FGC4 appears in [20]. However, the proof depends on a preprint of P. Vámos [19] for the nonreduced case, whereas Vámos in a note added to the preprint sent to me obtains the structure of σ-cyclic rings using a preprint of the Wiegands on the reduced case. (A beautiful example of international co-operation!) Brandal [27] has written a self-contained exposition.

A ring R is said to be right CF-cyclic, or FGCP, provided that every f. g. module M decomposes into a direct sum

$$M = R/I_1 \oplus R/I_2 \oplus \dots \oplus R/I_n$$

such that

$$I_1 \subseteq I_2 \subseteq \dots \subseteq I_n \subseteq \dots .$$

Clearly, every right CF-cyclic ring is right σ-cyclic. If R is commutative σ-cyclic, then by [19] and [20] R is CF-cyclic, and then the direct summands for the module M are called the canonical factors (CF's). Vámos' Theorem A of

[19] characterizes an FGCF ring R as an FSI Bezout ring, or equivalently, a σ-cyclic ring which is finitary in the sense that every ideal is an intersection of co-irreducible ideals. Every FSI ring is finitary, and a finite direct product $R^1 \times \ldots \times R^n$ of indecomposable FSI rings, each of which has a unique minimal prime P_i which is uniserial, and $R^i_{P_i}$ is AMVR. A local ring R is thereby FSI iff AMVR (and hence iff CFPF). Vámos' Theorem B states the indecomposable FSI rings are precisely the AMVR's or the locally maximal h-local domains or the locally almost maximal torch rings. (A ring R is a torch ring if R is locally almost maximal with at least two maximal ideals, a unique minimal prime P, and such that P is uniserial.)

18. Obvious Remark. Every commutative σ-cyclic ring is CFPF.

"Proof". Every factor ring is also σ-cyclic; hence, every f.g. R/A-module M is a direct sum of CF's, so if M is faithful over R/A, then R/A is one of the CF's.

19. Theorem. (Vámos). Any commutative σ-cyclic ring is quotient-injective, hence FSI.

Proof. Let R be an indecomposable σ-cyclic ring (FGCl-FGC4). Then for any maximal ideal M, the set of zero divisors is contained in M; hence, $Q = Q_{c\ell}(R)$ is a localization of R_M. But R_M is AMVR; hence Q is AMVR, hence injective, e.g., by 9D, so $Q = Q_{c\ell}(R) = Q_{max}(R)$. In the general case, $R = \prod_{i=1}^n R_i$, with $Q_{c\ell}(R) = \prod_{i=1}^n Q_{c\ell}(R_i) = Q_{max}(R)$ injective.

20. Corollary. (Vámos [19]). Every commutative σ-cyclic ring R is FSI, and $Q_{c\ell}(R/A)$ of every factor ring R/A is σ-cyclic.

Proof. Every factor ring R/A is σ-cyclic; hence the corollary follows from Theorem 19 and its proof which shows that $Q_{c\ell}(R/A)$ is a finite product of AMVR's; hence $Q_{c\ell}(R/A)$ is σ-cyclic.

21. Corollary. For any commutative Noetherian σ-cyclic ring R, $Q_{c\ell}(R)$ is QF, hence serial.

Proof. Follows from Corollary 20, since any Noetherian injective ring $Q_{c\ell}(R)$ is QF, hence a finite product of local QF Artinian rings, each of which is a VR. Thus, $Q_{c\ell}(R)$ is "serial".

The corollary is also a consequence of 18 and Theorem 11.

A ring R is said to be projective-free provided that all projective modules are free. In addition to local rings, these include principal ideal domains, and polynomial rings in finitely many variables over fields [28].

22. Proposition. Every σ-cyclic ring R is a finite product of projective-free rings.

Proof. Let R be one of the indecomposable σ-cyclic rings described by (FGC1)-(FGC4). Then, if P is any f. g. projective module, then P is a direct sum of cyclic modules R/A_i, $i = 1, \ldots, n$. But then R/A_i is projective, hence, A_i is a direct summand of R, hence indecomposability of $R \Rightarrow A_i = 0$, $i = 1, \ldots, n$, so $P \approx R^n$.

8. THE PRE-SIGMA CYCLIC CRITERION

In this section local FPF, PF, and FP^2F rings are characterized via the stated condition.

23. Theorem. A local ring R is right FPF iff for every integer $n > 0$, submodule K of R^n, we have the implication:

(23.1) $K \cap R^n a = 0$, $\forall a \in R \Longrightarrow \exists$ submodules A and B such that $B \supseteq K$, $A \approx R$, and $R^n = A \oplus B$.

In this case, $B \approx R^{n-1}$, and $M \approx R \oplus B/K$. (Moreover, in a right FP^2F local ring, every right ideal $\neq 0$ contains an ideal $\neq 0$.)

Proof. If R is right FPF, then since $M = R^n/K$ is a f.g. faithful module, then M generates mod-R; hence $M \approx R \oplus X$ for appropriate R-module X. Then $R^n = A_1 + B$ for submodules A_1 and B such that $A_1 \cap B = K$, $A_1/K \approx R$, $B/K \approx X$. Then, via projectivity of R, K splits in A_1, $K \oplus A = A_1$, so $R^n = K + A + B = A + B = A \oplus B$, and $B \supseteq K$ as stated. Since R is local, and $A \approx R$ and $B \approx R^{n-1}$ by Krull-Schmidt. (If R is right FP^2F, the same argument shows for a f.g. right ideal K that R/K faithful would imply $R/K \approx R \oplus B/K$, an impossibility unless B hence $K = 0$. Thus, every right ideal $K \neq 0$ contains an ideal $\neq 0$.)

Conversely, let M be any f.g. module. Then, $M \approx R^n/K$ for some submodule K of R^n, $n \geq 0$. Moreover, M is faithful implies that $K \cap R^n a = 0$, $\forall a \in R$, so (23.1) yields $R^n = A_1 \oplus B$ where $A_1 \approx R$ and $K \subseteq B \approx R^{n-1}$. Then $M = R^n/K \approx R \oplus B/K$ generates mod-R, and so R is right FPF.

Remark. Any semiperfect right FP^2F ring has strongly right bounded basic ring. (See [4], Theorem 3.1A or the proof of Theorem 1 of [171].)

24. Corollary. A local ring R is right FPF only if every faithful right module M generated by two elements is a direct sum of cyclics.

Proof. Write $M \approx R^2/K$. Then by Theorem 32, $M \approx R \oplus B/K$, where $B \approx R$, and the corollary follows.

25. Theorem. For a commutative local ring R, the f. a. e.:

(25.1) R is FPF.

(25.2) Every faithful module generated by two elements (or fewer) generates mod-R.

(25.3) Every faithful module generated by ≤ 2 elements is a direct sum of cyclics.

(25.4) Every submodule K of R^2 either embeds in a direct summand $\approx R$ or else $K \cap R^2 a \neq 0$ for some $a \in R$ (that is, R^2/K is not faithful).

Proof. Note that (2)<=> (4) by the proof of Theorem 23. Corollary 24 gives (1) => (3), and (3) => (2) via the proof of Corollary 24. Finally, assume (2). Then by the proof of 3E, R is uniform; hence (2) implies (as in the proof of Proposition 7) that every submodule M of $Q_c(R)$ generated by two elements is cyclic; hence every f.g. submodule is cyclic, so R is FPF by (2) of Theorem 9B. Thus (2) => (1).

Note: By requiring the modules to be finitely presented in (25.2-3) and by requiring K in (25.4) to be finitely generated, one obtains a characterization of FP^2F local rings. (Cf. Corollary 27.) Moreover, if one defines CFP^2F similarly to CFPF, then the next corollary can be modified to provide a characterization of CFP^2F local rings by the requirement that every finitely presented module generated by two elements is a direct sum of cyclics. But, these are simply the VR's. (See, e.g., [3b], p.130-131.)

26. Corollary. For a local ring R the f.a.e.:

(26.1) R is AMVR.

(26.2) R is CFPF.

(26.3) Every module generated by ≤ 2 elements is a direct sum of cyclics.

Proof. The equivalence of (1)-(3) already has been noted and (2) <==> (4) by the theorem.

As stated, a ring R is right PF provided that every faithful right R-module generates mod-R. (Consult [1], [3b], Chapter 42, [17] for background.)

Along the lines of Theorem 23 one can prove:

27. Corollary. Let R be any ring, and let $M = F/K$ be any right R-module, where F is a free right R-module. Then M generates mod-R iff there holds

(27.1) $\exists\, m > 0$ and submodules $A \approx R$ and $B \supseteq K^m$ such that $F^m = A \oplus B$.

In this case $M^m \approx R \oplus B/K^m$. (If R is local, then B is a free module.)

Moreover, R is right (F)PF iff for every (finitely generated) free right R-module F it is true that (27.1) holds for any submodule K such that $K \cap Fa = 0$, $\forall a \in R$, that is, F/K is faithful.

Similarly, R is right FP^2F iff the condition for right FPF holds with the proviso that K is finitely generated.

28. Further Problems (cf. Problem 15).

28A. In view of the Auslander-Bridger (stable) duality for f.p. modules over semiperfect rings (see, e.g., [3b], Chapter 25), one might conjecture that every semiperfect right FP^2F ring is also left FP^2F. However, the corresponding question is less promising for FPF, or even PF. (See [17] for the structure of the basic ring of a right PF ring.) Is every right PF ring left PF? By [17], it would be enough to prove that right PF implies left FPF!

28B. The structure theory for right PF rings has been determined by Azumaya [1], Osofsky, and Utumi (see Theorem G): R is right PF iff R is semiperfect right self-injective with essential right socle. (Then R is an injective cogenerator of mod-R, and conversely.) It would be interesting to have the right PF^2 rings characterized, that is, rings such that every faithful module F/K, with F free and K finitely generated, generates mod-R.

28C. A theorem of Cox [35] and Alamelu [36] states that if R is a commutative ring with Noetherian $Q_{c\ell}$, then $Q_{c\ell}$ is injective iff $End_R I$ is commutative, $\forall$ ideals I. In this case Q_{max} induces each $End_R I$ (as we remarked sup. prop. 1A). I conjecture the converse, namely, if Q induces each $End_R I$, $\forall$ ideals, then Q_{max} is injective. This would imply the Cox-Alamelu theorem. Incidentally, Baer's criterion states, essentially, that Q is injective iff Q induces I^* for all ideals!

28D. The determination of an FPF ring R ideal-theoretically is an unsolved problem except in special cases, e.g., when R is a domain; then R is FPF iff Prüfer. Also, CFPF is equivalent to R being AMVR, which can be expressed ideal-theoretically via the solvability of congruences

$$x \equiv x_{\underline{\alpha}} \bmod I_{\underline{\alpha}}$$

for any family of ideals $I_{\underline{\alpha}}$ and elements $x_{\underline{\alpha}}$, such that $\bigcap I_\alpha \neq 0$.

Local CFPF can be so described, and in view of the condition (25.4) which is just (25.3) applied to every factor ring, it appears likely that local FPF can be similarly expressed via congruences as a weakened form of AMVR, or linear compactness of R.

Note also that for $R = Q_{c\ell}$ this problem is equivalent (via 9D) to determining injectivity of R ideal-theoretically.

The condition CFPF = AMVR for local domain R has been determined by Matlis [23] via the condition that Q/R is injective. Thus, FPF $\Rightarrow$ Q injective, whereas Q/R injective $\Rightarrow$ CFPF.

The next theorem (which is not claimed to be new!) gives a characterization of situation (1) via commutativity of the endomorphism ring of the injective hull of R. (Cf. Lambek [24], p. 100, Exercises 2 and 3.)

29. Theorem. Let R be a commutative ring, let $E = E(R)$ be the injective hull, let $S = \mathrm{End}_R E$, and $Q = \mathrm{End}_S E$ (i. e., $Q = Q_{max}(R)$). Then $Q = \text{center } S$ canonically. Moreover, Q is injective (either as an R module, or as a Q-module) iff S is commutative.

Proof. (Consult Section 1 for background information.) Every element $x \in E$ can be written $x = s(1)$ for some $s \in S$, where 1 is the identity element of R; i.e., the mapping $r \mapsto xr$ of R onto xR is induced by an endomorphism s of E. Thus, $E = S1 = \{s(1) \mid s \in S\}$ is a cyclic left S-module, $E \approx S/\mathrm{ann}_S R$ (and $\mathrm{ann}_S R = \{s \in S \mid s(1) = 0\}$). It is known and easy to see that $Q = Q_{max}$ is commutative since given $f, g \in Q$; then there is a dense ideal H of R such that $\forall h \in H$, $fh \in R$, so $(fh)g = g(fh)$, so fg-gh annihilates the dense ideal H. Since $Q = \mathrm{End}_S E \subseteq \mathrm{End}_R E$ canonically, this means that fg-gh annihilates the maximal rational extension $\overline{R}$ of R in E; hence, fg-gh annihilates $Q = Q_{max} \approx \overline{R}$, that is; $fg = gh$, so Q is commutative. This implies, of course, that $Q \approx \text{center } S$ canonically.

Thus, if S is commutative, then $S = Q$, and $E = S1 = Q1 = \overline{R} \approx Q$ is an injective R-module, which implies Q is an injective Q-module and conversely. Next, if Q is injective, then $E = \overline{R} \approx Q$; hence, $S = \mathrm{End}_R E = \mathrm{End}_R Q \approx Q$ canonically, so S is commutative along with Q.

9. NOTE ON THE GENUS OF A MODULE AND GENERIC FAMILIES OF RINGS

Theorem 23 has been generalized in [34] to any product of FPF rings of genus 1, where the genus $g(R)$ of a ring is the infimum g of all integers γ such that for all f.g. generators M of mod-R, M^{γ} has a unimodular element; equivalently, there is an epic $M^{\gamma} \to R$. (If no such exists, then set $g = \infty$.) A family $F = \{R_i\}_{i \in I}$ of rings is generic of (with) bound B, if there exists a function $B : \mathbb{Z}^+ \to \mathbb{Z}^+$ such that for all modules M, if $\nu(M) < \infty$ is the minimal number of elements in any set of generators of M, then there is an epic $M^{B(\nu(M))} \to R$. The product theorem ([26]) states that any product of a generic family of rings of bound B is a ring which is generic of bound B (considering a ring as a family with one member). For example, a family of rings each of genus $\leq g$ is generic with bound $\leq g$, where g also denotes the constant function. Moreover, any family of commutative rings is generic of bound $1_{\mathbb{Z}^+}$. The 2×2 theorem of [34] states that if R is a commutative ring of genus 1, then for any faithful module M with $\nu(M) = 2$, the product M^2 has a unimodular element. Thus, by the product theorem the 2×2 theorem holds for any product of such rings.

A corollary of the product theorem is that any product $R = \prod_{i \in I} R_i$ of a generic family right FPF rings is right FPF. (In particular, the product of any family of commutative FPF rings is FPF.) This implies that any product of self-basic right FPF rings, in particular any product of self-basic right PF rings is right FPF.

Another corollary to the product theorem states that if $\{R_i\}_{i \in I}$ is any family of commutative rings, and if there exist integers $n > 0$ and $g > 0$ with the property that for all $i \in I$ every finitely generated R_i-module of free rank $\geq n+1$ has genus $\leq g$, then their product R has the same property: Every finitely generated R-module of free rank $\geq n+1$ has genus $\leq g$. The FPF theorem for commutative R is just the case $n = 0$ and $g = 1$.

ABBREVIATIONS

f. a. e. = following are equivalent.

f. g. = finitely generated.

f. p. = finitely presented.

PF = pseudo-Frobenius

FPF = finitely PF; FP^2F = finitely presented PF.

CFPF = completely FPF.

FSI = fractionally self-injective.

mod-R = the category of all (right) R-modules.

n. s. = nonsingular.

QF = quasi-Frobenius.

$Q_{c\ell}$ = the full, or classical, quotient ring.

Q_{max} = the maximal quotient ring.

rad-R = the Jacobson radical of R.

sing R = the singular ideal of R.

s. h. = semihereditary.

σ-cyclic = direct sum of cyclic modules.

σ-cyclic ring = one over which every f. g. module is σ-cyclic.

VR = valuation ring.

AMVR = almost maximal valuation ring.

w. r. t. = with respect to.

z. d. s. = zero divisors.

RESUMÉ OF THE MAIN UNSOLVED PROBLEMS

1. Does every (C)FPF ring R have injective $Q_{c\ell}$? Q_{max}? (Problem 15)

2. Is every right FPF Kasch ring right PF? (Cf. Remark 17.) Note: A semiperfect right FPF ring with nil radical is right PF [17I], hence right Kasch.

3. Is a right (F)PF also left (F)PF? Also consult Problem

4. Does every Noetherian FPF ring R have Robson splitting? (Cf. Theorems 10-11; also 15D.)

5. Does every pre-PF commutative ring R have Kasch $Q_{c\ell}$? (See Theorem 1G.)

6. Determine ideal-theoretically all FPF rings, CFPF rings, FP^2F rings, CFP^2F rings.

7. Let g(R) be modified by deleting the f.g. condition imposed in generators, and and let G(R) be the resulting integer, called the big genus of R., that is, the genus for "big" modules. Obviously $g(R) \leq G(R)$. How may they differ?

8. Every 2-sided FPF prime ring is 2-sided Goldie [4], and every right Goldie right FPF prime ring is 2-sided Goldie [4] so the question is: Is a right FPF prime ring right (left) Goldie?

9. Is a right Noetherian right FPF prime ring also left Noetherian? (See [4] for partial results.)

10. If every ideal has commutative endomorphism ring, does R have injective Q_{max}? As stated, the converse is true. See Problem 28D. Also note that every faithful ideal of any commutative ring has commutative endomorphism ring. (Prop. 1B.)

REFERENCES

[1] Azumaya, G., Completely faithful modules and self-injective rings, Nagoya J. Math., 27 (1966), 249-278.

[2] Cateforis, V. C., Flat regular quotient rings, Trans. Amer. Math. Soc., 138 (1969), 241-249.

[3a] Faith, C., Algebra: Rings, Modules and Categories I, Grundlehren der Math. Wiss., 190, Springer-Verlag, Berlin-Heidelberg-New York, 1973.

[3b] ________, Algebra II: Ring Theory, Grundlehren der Math. Wiss., 191, Springer-Verlag, Berlin-Heidelberg-New York, 1976.

[4] ________, Azumaya-Morita Theory: Faithful Modules and Generators of Mod-R, in preparation.

[5] Faith, C., and Zaks, A., Injective quotient rings, preprint, Dept. of Math., Rutgers Univ., New Brunswick, N. J., 1976.

[6] Goodearl, K. R., Embedding non-singular modules in free modules, J. Pure and Appl. Algebra, 1 (1971), 275-279.

[7] Goldie, A. W., Semiprime rings with maximum condition, Proc. London Math. Soc., X (1960), 201-220.

[8] Popescu, N., and Spircu, T., Quelques observations sur les epimorphismes plat (à gauche) d'anneaux, J. Algebra, 16 (1970), 40-59.

[9] Sandomierski, F. L., Nonsingular rings, Proc. Amer. Math. Soc., 19 (1968), 225-230.

[10] Silver, L., Noncommutative localizations and applications, J. Algebra, 7 (1967), 44-76.

[11] Stenstrom, B., Rings of Quotients, Grundl. der Math. Wiss, 217, Springer-Verlag, Berlin-Heidelberg-New York, 1975.

[12] Storrer, H. H., Epimorphic extensions of non-commutative rings, Comment. Math. Helv., 48 (1973), 72-86.

[13] ____________, A characterization of Prüfer domains, Canad. Math. Bull., 12 (1969), 809-812.

[14] Robson, J. C., Decompositions of Noetherian modules, Comm. Algebra, 4 (1974), 345-349.

[15] Tachikawa, H., A generalization of quasi-Frobenius rings, Proc. Amer. Math. Soc., 20 (1969), 471-476.

[16] Faith, C., Semiperfect Prüfer and FPF rings, Israel Math. J., 26 (1976), 166-177.

[17] ________, Injective cogenerator rings and a theorem of Tachikawa, I, II, Proc. Amer. Math. Soc., 60 (1976), 25-30; 62 (1977), 15-18.

[18] ________, Self-injective rings, Proc. Amer. Math. Soc., (1978).

[19] Vámos, P., The decomposition of finitely generated modules and fractionally selfinjective rings, preprint, Univ. of Sheffield, Sheffield, England, 1977.

[20] Wiegand, R., and Wiegand, S., Commutative rings over which finitely generated modules are direct sums of cyclics, preprint, Univ. of Nebraska, Lincoln, Neb. 68588, 1977.

[21] Bass, H., On the ubiquity of Gorenstein rings, Math. Z., 82 (1963), 8-28.

[22] Matlis, E., Injective modules over Noetherian rings, Pac. J. Math., 81 (1958), 511-528.

[23] ________, Injective modules over Prüfer rings, Nagoya Math. J., 15 (1959), 57-69.

[24] Lambek, J., Rings and Modules, Blaisdell, New York, 1966; corrected reprint, Chelsea, 1976.

[25] Endo, S., Completely faithful modules and quasi-Frobenius algebras, J. Math. Soc. Japan, 19 (1967), 437-456.

[26] Utumi, Y., On rings of which any one-sided quotient rings are two-sided, Proc. Amer. Math. Soc., 14 (1963), 141-147.

[27] Brandal, W., Commutative rings whose finitely generated modules decompose, preprint, Univ. of Tennessee, Knoxville, 1977.

[28] Quillen, D. R., Projective modules over polynomial rings, Invent. Math., 36 (1976), 167-171.

[29] Kaplansky, I., Rings of Operators, Benjamin, New York and Amsterdam, 1968.

[30] Chase, S. U., Direct products of modules, Trans. Amer. Math. Soc., 97 (1960), 457-473.

[31] Kaplansky, I., Elementary divisors and modules, Trans. Amer. Math. Soc., 66 (1949), 464-491.

[32] Warfield, R. B., Jr., Decomposability of finitely presented modules, Proc. Amer. Math. Soc., 25 (1970), 167-172.

[33] Utumi, Y., On continuous regular rings and semisimple self-injective rings, Canad. J. Math., 12, 597-605.

[34] Faith, C., The genus of a module and generic families of rings, in preparation.

[35] Cox, S. H., Jr., Commutative endomorphism rings, Pac. J. Math., 45 (1973), 87-91.

[36] Alamelu, S., On commutativity of endomorphism rings of ideals, Proc. Amer. Math. Soc., 37 (1973), 29-31.

INDEX

SPECTRUM, TOPOLOGIES AND SHEAVES FOR LEFT NOETHERIAN RINGS

By Zoltan Papp

George Mason University, Fairfax, Virginia 22030 USA

1. Introduction. The problem of representing a ring by the global sections of a sheaf can be formulated as follows. Given a ring R, find a topological space X and a sheaf $\tilde{R}$ of rings over X such that $\Gamma(X,\tilde{R}) \cong R$, i.e. R is isomorphic to the ring of continuous sections over X. A well known example of this is the structure sheaf over Spec R, the collection of prime ideals with the Zariski topology, for a commutative ring R where the technique of localizations is the main tool used in the constructions. Our main problem is to find those noetherian rings R for which the representation problem has a solution and the structure sheaf can be constructed using the localizations of rings and modules. The main result of this paper shows that this representation is possible when R is a left stable, left noetherian ring. A ring R is called left stable if for every torsion theory τ and R-module M, M is τ-torsion exactly when E(M) is τ-torsion. E(M) is the injective envelope of M.

By introducing the following concepts we can make our problem more exact. (For any additional undefined concept we refer to the book [1] of J. Golan.)

There are three basic candidates for the spectrum of a ring R.

1. Spec R = {the collection of prime ideals of R}
2. R-sp = {the collection of prime torsion theories on R-mod}
3. Sp(R-mod) = {the collection of isomorphism classes of indecomposable injective R-modules}.

Since the technique of localizations of noncommutative rings is based on the concept of injective modules, and the correspondence between prime ideals and the isomorphism classes of indecomposable injective R-modules is not one-to-one in general, Spec R has a disadvantage in the applications.

On the other hand, if R is a left noetherian ring, then the assignment $\Psi: F \mapsto \chi(F)$, where $F \in$ Sp(R-mod) and $\chi(F)$ is the largest torsion theory for which F is torsion free, also $\chi(F) \in R$-sp, is a bijection between the sets Sp(R-mod) and R-sp. This makes possible to identify them, and the two sets thus identified will be denoted by X and, with an appropriate topology T, it will be called the spectrum of the ring R. (We are going to use X for Sp(R-mod) or R-sp depending on which gives an advantage in expressing our concepts.)

Given any topology T on X and an R-module M a presheaf can be defined on X as follows. For an open subset U of X $\wedge U$ denotes the meet of all prime torsion theories in U, i.e. $\wedge U = \wedge\{\pi \mid \pi \in U\}$ or, alternatively, $\wedge U = \wedge\{\chi(F) \mid F \in U\} = \chi(\oplus\{F \mid F \in U\})$ where we used the fact that Sp(R-mod) and R-sp are identified by the bijection Ψ: $F \mapsto \chi(F)$ and F denotes both an isomorphism class of indecomposable injective modules and one of its representatives. If $M \in$ R-mod, then the assignment $\tilde{M}$: $U \mapsto Q_{\wedge U}(M)$ gives a separated presheaf on X which will be called the Q-presheaf $\tilde{M}$ on X for the given topology T. (For detail see the book [1] of J. Golan.) The stalk of the presheaf $\tilde{M}$ at an element $\pi \in X$ is defined as usual, $S_\pi(\tilde{M}) = \varinjlim_{\pi \in U} Q_{\wedge U}(M)$, and there exists a canonical homomorphism $\varphi_\pi(M)$: $S_\pi(\tilde{M}) \to Q_\pi(M)$ where $Q_\pi(M)$ is the localization of M at the prime torsion theory π. Our problem can now be formulated in the following way.

Main problem: Find all those noetherian rings R for which there exists a topology T on X such that the Q-presheaf $\tilde{R}$ is a sheaf and $S_\pi(\tilde{R}) \cong Q_\pi(R)$. In this case R can be represented as a ring of continuous sections over X with values in the quotient rings $Q_\pi(R)$, $\pi \in X$. Similar question can be asked about the R-modules M.

One class of left noetherian rings where all the above requirements are satisfied is the class of left stable, left noetherian rings. This result comes from the following sources.

A. The construction of sheaves on Sp(R-mod) by B. Goldston and A. C. Mewborn in [3].

B. In papers [2] and [6] J. Golan, J. Raynaud and F. Van Oystaeyen constructed presheaves on R-sp with good local properties.

C. A topological characterization of the left stable, left noetherian rings in [5] by the author.

2. The Goldston-Mewborn sheaves for left noetherian rings.

Let $X = Sp(R\text{-mod})$. Given $F_1, F_2 \in X$ we say that there is a directed edge from F_1 to F_2 if $F_1 \cong F_2$ or there exists a submodule A of F_1, a cocritical submodule S_1 of A and a cocritical submodule S_2 of F_2 such that $0 \to S_1 \to A \to S_2 \to 0$. If two elements $F, G \in X$ are connected by a finite sequence of directed edges, we say that there is a directed path from F to G. Given a left ideal I of R, let $C(I) = \{G \mid G \in X$ and there exists $F \in X$ such that $\mathrm{Hom}(R/I, F) \neq 0$ and there is a directed path from F to $G\}$. Define $\mathcal{O}(I) = X \setminus C(I)$, it was shown in [3] that the set $\{\mathcal{O}(I) \mid I$ is left ideal of $R\}$ is a base of open sets for a topology on X that will be called the Goldston-Mewborn topology on X. B. Goldston and A. C. Mewborn proved the following theorem in their paper [3].

Theorem A. Let R be a left noetherian ring. If M is an R-module with finite uniform dimension or $X = Sp(R\text{-mod})$ is a noetherian space, then the Q-presheaf $\tilde{M}$ is a sheaf for the R-module M. In particular $\tilde{R}$ is a sheaf of rings. □

The following result points out an interesting property of the Goldston-Mewborn topology of X and it is also one of the reasons why the Q-presheaves are sheaves for this topology.

Proposition 1. Let R be a left noetherian ring, U an open subset of X and let F be an indecomposable injective

R-module. Then F is $\wedge U$-torsion if and only if $F \notin U$. Moreover $\wedge U$ is a stable torsion theory for every open subset U of X.

Proof. Since $\wedge U = \wedge\{\chi(G) | G \in U\} = \chi(\oplus\{G | G \in U\})$, $F \in U$ clearly implies that F is $\wedge U$-torsion free. Assume that $F \notin U$. If F is not $\wedge U$-torsion, then $\mathrm{Hom}(F, \oplus\{G | G \in U\}) \neq 0$ which shows that for some $G_o \in U$, $\mathrm{Hom}(F, G_o) \neq 0$. Thus, by Lemma 2.1 of [3], there exists a directed path from F to G_o, hence $G_o \in \mathrm{cl}(\{F\})$, where $\mathrm{cl}(V)$ denotes the closure of the subset V of X in the given topology. (See the proof of Theorem 2 in [4].) Since $F \in \bar{U} = X \setminus U$ is closed it follows that $G_o \in \mathrm{cl}(\{F\}) \subseteq \bar{U}$ which is a contradiction. This proves that F is $\wedge U$-torsion. We have seen that for an open subset U of X and for any indecomposable injective module F, F is either $\wedge U$-torsion free or $\wedge U$-torsion module. In case of a left noetherian ring this insures that $\wedge U$ is a stable torsion theory. □

Unfortunately the Goldston-Mewborn sheaf is rather coarse, and the map $\varphi_\pi(M) : S_\pi(\tilde{M}) \to Q_\pi(M)$ is not monomorphism in general. This follows from the following results.

Proposition 2: Let R be a left noetherian ring, $\mathcal{T}$ a topology on $X = \mathrm{Sp}(R\text{-mod})$, $M \in R\text{-mod}$ and consider the Q-presheaf $\tilde{M}$ on X. Then each of the following conditions implies the next.

(a) $\varphi_{\chi(F)}(M) : S_{\chi(F)}(\tilde{M}) \to Q_{\chi(F)}(M)$ is a monomorphism for all $M \in R\text{-mod}$ and $F \in X$.

(b) If $G \in V_F = \cap\{U | U$ open in $\mathcal{T}$ and $F \in U\}$, then G is $\chi(F)$-torsion free.

(c) The space X with the topology T is a T_0-space.

Proof. Let T be a topology on X and assume that (c) is not true. Then there exist $F,G \in X$, $F \not\cong G$ such that $cl(\{F\}) = cl(\{G\})$. This means that for every open set $U, G \in U$ if and only if $F \in U$, hence $F \in V_G$ and $G \in V_F$. If condition (b) is true, then G is $\chi(F)$-torsion free and F is $\chi(G)$-torsion free, hence $\chi(F) = \chi(G)$ follows. Since R is a noetherian ring we have $F \cong G$ which is a contradiction. Assume now that (b) is not true, then there exist $F,G \in X$ such that $G \in V_F$ but G is not $\chi(F)$-torsion free. Let U be any open neighborhood of F, then $G \in U$ and $Q_{\wedge U}(G) = G$, hence $S_{\chi(F)}(\tilde{G}) = \varinjlim_{F \in U} Q_{\wedge U}(G) = G$. Since $G = S_{\chi(F)}(\tilde{G})$ is not $\chi(F)$-torsion free, the map $\varphi_{\chi(F)}(G)\colon S_{\chi(F)}(\tilde{G}) \to Q_{\chi(F)}(G)$ cannot be a monomorphism. □

Proposition 3. Let R be a left noetherian ring. Consider the Goldston-Mewborn sheaves on $X = Sp(R\text{-mod})$ and for $M \in R\text{-mod}$ let $\varphi_{\chi(F)}(M)\colon S_{\chi(F)}(\tilde{M}) \to Q_{\chi(F)}(M)$ be the canonical homomorphism. Then $\varphi_{\chi(F)}(M)$ is a monomorphism for every $F \in X$ and $M \in R\text{-mod}$ with finite uniform dimension if and only if R is a left stable ring.

Proof. The proof of the "if" part will follow from Theorem 1. Assume that $\varphi_{\chi(F)}(M)$ is a monomorphism and let $F,G \in X$ such that $Hom(G,F) \neq 0$. Lemma 2.1 of [3] implies that there exists a directed path from G to F, thus from the definition of the closed sets in the Goldston-Mewborn topology we have that $F \in cl(\{G\})$ from which $G \in V_F$ follows. By Proposition 2 G

is $\chi(F)$-torsion free, i.e. $\chi(F) \leq \chi(G)$. An application of Theorem of [4] shows that R is stable. □

3. <u>The Golan-Raynaud-Van Oystaeyen presheaf</u>. Given a ring R, let X = R-sp and call an element $\pi \in$ R-sp the prime generalization of a torsion theory τ if $\tau \leq \pi$. The set $\{p \text{ gen } [\xi(R/I)] \mid I$ is left ideal of $R\}$ forms a base of open sets for a topology on X that is called the basic order topology on X. In the papers [2] and [6] J. Golan, J. Raynaud and F. Van Oystaeyen presented the following results.

<u>Theorem B</u>. Let R be a left noetherian ring. Then the Q-presheaf on X = R-sp with the basic order topology has the following properties.

(a) $\varphi_\pi(M): S_\pi(\tilde{M}) \to Q_\pi(M)$ is monomorphism for every R-module M and $\pi \in X$.

(b) If M is finitely generated, then $\varphi_\pi(M)$ is an isomorphism. □

Together with Proposition 2 this implies the following corollary.

<u>Corollary</u>. Given a ring R, then R-sp with the basic order topology is a T_0-space. □

Theorem B (b) can be extended to any R-module M.

<u>Proposition 4</u>. Let R be a left noetherian ring. Then the map $\varphi_\pi(M): S_\pi(\tilde{M}) \to Q_\pi(M)$ is an isomorphism for every $M \in$ R-mod and $\pi \in X$.

Proof. Let M be a module over the noetherian ring R. Then the set of finitely generated submodules of M forms a directed system $\{M_k | k \in K\}$ where $k \leq j$ is equivalent to $M_k \subseteq M_j$ and $\varphi_{kj}: M_k \to M_j$ is the inclusion map. It also follows that $M = \varinjlim_{k\in K} M_k$. Since R is a noetherian ring, Q_τ commutes with direct limits for every torsion theory τ. Theorem B (b) gives the isomorphism $\varinjlim_{\pi\in U} Q_{\wedge U}(N) \cong Q_\pi(N)$ for every finitely generated R-module N and we also have that direct limits commute. The applications of the preceding statements give the following isomorphisms

$$Q_\pi(M) \cong Q_\pi(\varinjlim_{k\in K} M_k) \cong \varinjlim_{k\in K} Q_\pi(M_k) \cong \varinjlim_{k\in K} (\varinjlim_{\pi\in U} Q_{\wedge U}M_k)) \cong$$

$$\varinjlim_{\pi\in U} \varinjlim_{k\in K} Q_{\wedge U}(M_k)) \cong \varinjlim_{\pi\in U} Q_{\wedge U}(\varinjlim_{k\in K} M_k) \cong \varinjlim_{\pi\in U} Q_{\wedge U}(M) = S_\pi(\tilde{M})$$

which prove our result. □

4. Structure sheaves for left stable, left noetherian rings.

Let R be a commutative noetherian ring. Then we have the identifications $X = \text{Spec } R = R\text{-sp} = \text{Sp}(R\text{-mod})$ through bijections and both the basic order topology and the Goldston-Mewborn topology coincide with the Zariski topology on X. Also the Goldston-Mewborn sheaf and the Golan-Raynaud-Van Oystaeyen presheaf becomes the usual structure sheaf of the ring R. Since commutative noetherian rings are stable, we can ask the question about the relationship of the basic order topology and the Goldston-Mewborn topology on X and their respective presheaves if R is any left stable, left noetherian ring. The following theorem of [5] gives the answer to the question about the topologies.

<u>Theorem C</u>. Let R be a left noetherian ring and use the bijection $\Psi: F \mapsto \chi(F)$ for the identification $X = Sp(R\text{-mod}) = R\text{-sp}$. Then R is a left stable ring if and only if the Goldston-Mewborn topology coincides with the basic order topology on X. □

The equality of the topologies implies that the Q-presheaves are the same as well, thus we have the following theorem which gives a partial answer to our main problem.

<u>Theorem 1</u>. Let R be a left stable, left noetherian ring and X is the spectrum of R. (i.e. $X = R\text{-sp} = Sp(R\text{-mod})$ with the basic order topology which is the same as the Goldston-Mewborn topology.) If M is an R-module with finite uniform dimension, then

(a) the Q-presheaf $\tilde{M}$ is a sheaf,

(b) the map $\varphi_\pi(M): S_\pi(\tilde{M}) \to Q_\pi(M)$ is an isomorphism for every $\pi \in X$,

(c) $\Gamma(X,\tilde{M}) \cong M$ and $\Gamma(X,\tilde{R}) \cong R$. □

This result provides a solution for the representation problem in the case when R is a left stable, left noetherian ring.

Remark. The above theorem proves the "if" part of Proposition 3.

Let us end our discussions with the following problems.

1. Let R be a left stable, left noetherian ring. Is X, the spectrum of R, a noetherian space? The affirmative answer would imply, via Theorem A, that in our Theorem 1 we would not have to assume that M has finite uniform dimension.

2. Proposition 3 characterizes those rings for which the Goldston-Mewborn sheaf "behaves well", i.e. $\varphi_{X(F)}$ is an isomorphism for any $F \in X$. On the other hand, the stalks of the Golan-Raynaud-Van Oystaeyen presheaf "behaves well", but $\tilde{M}$: $U \mapsto Q_{\wedge U}(M)$ is only a presheaf. The problem is how to characterize those (left noetherian) rings for which the Q-presheaf on X with the basic order topology is a sheaf. (This is a part of our main problem.)

REFERENCES

[1] J. Golan, Localization of Noncommutative Rings, Marcel Dekker Inc., New York, 1975.

[2] J. Golan, J. Raynaud and F. Van Oystaeyen, Sheaves over the spectra of certain noncommutative rings, Comm. Algebra, 4 (1976) 491-502.

[3] B. Goldston and A. C. Mewborn, A structure sheaf for a noncommutative noetherian ring, to appear in J. Algebra.

[4] Z. Papp, On stable noetherian rings, Trans. Amer. Math. Soc. 213 (1975), 107-114.

[5] Z. Papp, A topological characterization of stable rings, Arch. Math. 29 (1977) 235-240.

[6] F. Van Oystaeyen, Stalks of sheaves over the spectra of certain noncommutative rings, to appear in Comm. Algebra.

PROBLEMS

FULLY LEFT BOUNDED LEFT NOETHERIAN RINGS

John Beachy

A left Noetherian ring R is <u>fully left bounded</u> $\Longleftrightarrow$ for each cyclic module $_RM$ there exist elements $m_1,\ldots,m_n \in M$ such that $\mathrm{Ann}(M) = \mathrm{Ann}(m_1,\ldots,m_n)$. Question: For what rings is there a uniform bound on the number of elements required? In particular, does a Noetherian ring with polynomial identity have a uniform bound? The condition is easily seen to be satisfied for any left Artinian ring or any ring finitely generated (as a module) over its center. A left Noetherian ring has bound one if and only if every left ideal is two-sided.

BOUNDED PRIME RINGS, PSEUDO-FROBENIUS RINGS, THE JACOBSON RADICAL OF A RING

Carl Faith

In this note I discuss the following two problems:

1. Does projectivity of the (f.g.) ideals of a bounded prime ring R imply that R is semihereditary?

2. When is a right PF ring also left PF, e.g., always?

The conjectures of Jacobson, briefly, are on the structure of the Jacobson radical J of R under the conditions:

JC 1. If R is Noetherian, is $\bigcap_{n<\infty} J^n = 0$

JC 2. If R is a f.g. algebra, is J nil?

JC 3. If R = FG is the group ring of a group over a field F of characteristic 0, is J = 0?

In greater detail:

1. <u>A Problem on Prime Rings with Projective Ideals.</u>

A. Let R be a right bounded prime ring such that every finitely generated ideal of R is projective in the category mod-R of all right R-modules. Then is R right semihereditary?

By the Asano-Michler Theorem, the answer is "yes" when R is Noetherian (Michler, Proc. LMS (1969)). For shorter proofs, see Griffith-Robson, PAMS (1970), and Lenagan, Bull. LMS (1971).

A theorem of Faith, Israel Math. J. (1976), proves it for semiperfect ring R, in which case $R \approx D_n$ is a full $n \times n$ matrix ring over right and left serial domain D. (Warfield has pointed out that the proof shows that any semiperfect ring satisfying the hypothesis that f.g. ideals are projective on the right is semihereditary, and, moreover, if R is prime, then $R = D_n$ as above, with D a right duo ring.)

B. Same question assuming that all ideals are projective in mod-R.

C. Same question assuming that (f. g.) ideals are projective on both sides.

D. Same question assuming that every (f. g.) ideal $\neq 0$ is an f. g. projective generator in mod-R.

E. Same question assuming that every (f. g.) ideal is invertible in the Goldie quotient ring Q, that is, that every (f. g.) ideal is f. g. projective and a generator on both sides.

F. Same question assuming that every (f. g.) ideal is a principal right ideal $\approx R$ (or even a principal right and left ideal $\approx R$ on both sides).

Remark. If R is semiperfect, and if every f. g. ideal $\neq 0$ of R generates mod-R, then $R \approx D_n$, where D is a local semifir. For this implies that every f. g. right ideal generates mod-R, and then the proof of the result of Faith cited above suffices for this. (In this connection, see a paper of Cohn, Nagoya (1966), on the structure of a local semifir.)

2. A Problem on Pseudo-Frobenius Rings: Is a right injective cogenerator ring also left injective?

A ring R is said to be right Pseudo-Frobenius (PF) if R satisfies the following equivalent conditions:

PF1. R is a right self-injective semiperfect ring with essential right socle.

PF2. R is right self-injective with finite essential right socle.

PF3. R is an injective cogenerator of mod-R.

PF4. R is injective and every simple right module embeds in R.

PF5. R is semilocal and cogenerates mod-R.

PF6. $R = \oplus_{i=1}^{n} e_i R$, where $e_i = e_i^2 \in R$ is such that $e_i R e_i$ is a local ring, and $e_i R$ is injective with simple right socle, $i = 1, \ldots, n$.

Thus, these are apparently nice rings. However, the following question has been open for some time:

A. Is every right PF also left PF?

See Faith (PAMS, 1976, 1977), for background references and partial results. These papers also contain new results on the structure of right PF rings, e.g., the basic ring is strongly right bounded in the sense that every right ideal $\neq 0$ contains an ideal $\neq 0$. A theorem of Kato asserts that for right PF to imply left PF it suffices to prove that right PF implies that R is left self-injective (Proc. J. Acad., 1968).

The main result of my 1978 PAMS paper on self-injective rings is the following:

2. Theorem. *Let* $R = (B, E)$ *be the semidirect product of a bimodule* E *over a ring* B. *Thus,* $a(xb) = (ax)b$ *for all* $a, b \in B$ *and* $x \in E$, *and in* $R = B \times E$ *addition is componentwise, and multiplication is defined by:*

(2.1) $$(a, x)(b, y) = (ab, ay+xb).$$

Then:

(2.2) R *is right self-injective iff* E *is injective in* mod-B, *and* $B = \operatorname{End} E_B$ *canonically.*

(2.3) R *is a right injective cogenerator in* mod-R (= R *is right* PF) *iff* E *is an injective cogenerator of* mod-B *satisfying* $B = \operatorname{End} E_B$ *canonically.*

(2.4) Assuming (2.3), then R is left PF iff E is an injective cogenerator of B-mod, and $B = \text{End}_B E$ canonically.

The cogenerator case is a special case of a result of Müller, Can. J. Math., 1969.

The proof of Theorem 2 depends on the following lemma.

1. Lemma. Let R be a ring, let E be an ideal which is its own left annihilator, ${}^{\perp}E = \{a \in R \mid aE = 0\} = E$, let $B = R/E$. Then E is canonically a B-bimodule. If

(1.1) E is injective as a (canonical) right B-module, and

(1.2) $B \approx \text{End}\, E_B$ canonically,

then, R is right self-injective.

Conversely, if R is right self-injective, then for any ideal A, the left annihilator ${}^{\perp}A$ is an injective right R/A-module, and $\text{End}\,{}^{\perp}A_{R/A} \approx R/{}^{\perp\perp}A$ canonically. Thus, in this case, any ideal E satisfying $E = {}^{\perp}E$ satisfies (1.1) and (1.2).

Note: The proof of the lemma is direct in the sense that one shows that the injective hull F of R must equal R; that is, one does not use Baer's criterion for injectivity.

2B. Corollary. If every right PF ring is left PF, then a bimodule E over a ring B satisfies (2.3) iff it satisfies the left-right symmetry (2.3)'.

In view of (2.4) of Theorem 2, Question 2 therefore has a negative answer if the answer to the following question is affirmative.

A. Does there exist a (B, B)-bimodule E which is an injective cogenerator in mod-B such that $B = \text{End}\, E_B$ but the left-right symmetry (2.4) fails to hold? For then R = (B, E) is right but not left PF.

A related question:

B. If E is injective in mod-B and $B = \text{End}\, E_B$ canonically, is then E injective in B-mod, that is, as a left B-module? And if so, is $B \approx \text{End}_B E$ the endomorphism ring of the left B-module E?

See my paper (PAMS, 1978) for a discussion of Corollary 2B.

The main advantage of Corollary 2.3 is that even if right PF does not imply left PF, nevertheless it may be true that (2.3) $\Rightarrow$ (2.3)' for various classes of rings, e.g., B an integral domain.

THREE CONJECTURES OF JACOBSON

JC1. If R is any 2-sided Noetherian ring and J is the Jacobson radical, does the Krull Intersection Theorem hold: $\bigcap J^n = 0$.

Herstein showed the answer is "no" if R is just right Noetherian (A counterexample in Noetherian rings, Proc. N. A. S. (1965)). Jategaonkar (J. Algebra (1974)) and independently Cauchon (Comm. Alg. (1976)) affirmed the conjecture for fully bounded Noetherian (FBN) rings. Lenagan proved it for Noetherian rings of Kdim = 1 (JLMS 1977).

JC2. Let R be an algebra over a field F finitely generated as an algebra (i.e., a homomorphic image of the free algebra $F[X]$ on a finite set X). Is J a nil ideal? As reported in Jacobson's revised colloquium volume (in one of the appendices), this was solved independently by Goldman and Krull (1950) for commutative R, and by Amitsur (PAMS, 1960) assuming F is non-denumerable.

JC 3. If $R = FG$, a group ring over F of characteristic 0, then is $\text{Rad}\, R = 0$? Amitsur (same paper) showed the answer is "yes" if F is non-denumerable. D. G. Passman has written a great deal on this difficult problem (Infinite Group Rings, Dekker, 1971; also see his report in the 1974 Israel Math. Journal).

Regarding these two problems, one might try to solve it using recent techniques of Cauchon and Jategaonkar on their solution to Jacobson's conjecture for fully bounded Noetherian (FBN) rings, i.e., do Jacobson's Conjectures 2 and 3 hold for FBN algebras finitely generated over F?

COMMUTATIVE NOETHERIAN LOCAL RINGS

Melvin Hochster

Let $(R,\mathfrak{m})$ be a commutative Noetherian local ring with identity, Krull dim $R = n$ and $x_1,\ldots,x_n$ a system of parameters. The following questions are open if dim $R \geq 3$ and R has mixed characteristic.

1. If T is finitely generated, $T \neq 0$, $\mathrm{id}_R T < \infty$, is R Cohen-Macaulay?

2. If M is f.g., $M \neq 0$, pd $M < \infty$, and $x \notin \mathrm{zd}(M)$, is $x \notin \mathrm{zd}(R)$? ($\mathrm{zd}(M) = \{a \in R \mid ax = 0$ for some $0 \neq x \in M\}$.)

3. With M as in 2), $I = \mathrm{Ann}_R M$, $h: R \to S$, S Noetherian, Q a minimal prime of IS, is dim $S_Q \leq \mathrm{pd}_R M$?

4. If $0 \to F_d \to \ldots \to F_0 \to 0$ is a finite free complex with $H_o(F_o) \neq 0$ of finite length, is dim $R \leq d$?

5. If $S \subset R$, S regular local, R module-finite over S, is S a direct summand of R as an S-module?

6. If R is a domain, M a finitely generated torsion-free R-module, $u \in \mathfrak{m}M$, $\mathrm{Tr}\, u = \{\phi(u) : \phi \in \mathrm{Hom}_R(M,R)\}$ and Q is a minimal prime of $\mathrm{Tr}\, u$, is dim $R_Q \leq$ rk M?

7. (a) Is there a (not necessarily finitely generated) R-module E such that $(x_1,\ldots,x_n)E \neq E$ and $x_1,\ldots,x_n$ is an E-sequence? (b) If R is complete, is there a finitely generated such E?

<u>Remarks</u>. 7b $\Longrightarrow$ 7a $\Longrightarrow$ {4),5),6)} while 4) $\Longrightarrow$ 3) $\Longrightarrow$ {1),2)}. 7a) is known if R contains a field, while 7ab) are known if dim R $\leq$ 2. All these questions are open if R has mixed characteristic and dim R $\geq$ 3. 7b) is open in almost all cases if dim R $\geq$ 3.

CONTINUOUS AND DUAL-CONTINUOUS MODULES

Saad Mohamed

A module is called local if it has a proper submodule which contains every proper submodule.

A module M is called continuous if it satisfies the following:

(I) Every submodule of M is essential in some summand of M.

(II) For every summand M' of M, every exact sequence $0 \to M' \to M$ splits.

A module M is called dual continuous (d-continuous) if it satisfies the following:

(I) For every submodule A of M, M decomposes as $M = M_1 \oplus M_2$ where $M_1 \subset A$ and $A \cap M_2 \subset_s M$.

(II) For every summand M' of M, every exact sequence $M \to M' \to 0$ splits.

[Note: $A \subset_s B$ reads A is small in B.]

1. Ahsan (1973) proved that a ring R all of whose cyclic modules are quasi-injective is semi-perfect. Is it true that a ring R all of whose cyclic modules are continuous is also semi-perfect? It is enough to consider the case when R is a regular ring.

2. A ring R is (semi-)perfect if and only if every (finitely

generated) quasi-projective R-module is d-continuous. Characterise those rings for which every d-continuous R-module is quasi-projective.

3. A d-continuous module M has a unique (up to isomorphisms) decomposition into the direct sum of d-continuous modules,

$M = (\sum_{i \in I} \oplus A_i) \oplus M'$ where A_i is a local module and Rad $M' = M'$.

What is the structure of M'?

What are the conditions which make such a direct sum d-continuous?

LEFT STABLE LEFT NOETHERIAN RINGS

Z. Papp

Let R be a left stable left noetherian ring. Consider X = R-sp (the left spectrum of R) with the basic order topology. (For definitions, see Golan, Localization of Non-commutative rings, Marcel Dekker, Inc., N.Y. 1975.)

Question: Is X a Noetherian space?

FINITELY GENERATED ALGEBRAS OVER A FIELD

Martha Smith

1) Let R be a finitely generated algebra over F such that every irreducible R-module M has $\mathrm{End}_R M$ algebraic over F. What can you say about R ring-theoretically?

2) Let D be a division algebra over F which is finitely generated. Is $[D:F] < \infty$? If F is non-denumerable, then D is algebraic over F.

SIMPLE NOETHERIAN RINGS

Toby Stafford

In the lecture on 'A simple Noetherian ring not Morita equivalent to a domain', I showed that the following results held for a simple Noetherian **ring R with KdimR=1**

a) Given d essential in R, then there exists f in R such that $R = dR + fdR$.

b) If M is a fin. gen. torsion R-module, then M is cyclic.

c) If I is a right ideal of R, then I is two-generated.

d) If M is a fin. gen. torsion-free R-module, then $M \cong I \oplus R^{(s)}$ for some right ideal I of R.

I would like to know to what extent these results hold for arbitrary simple Noetherian rings. Certainly all these results do have a generalisation to rings of higher Krull dimension. For example, a right ideal of a simple Noetherian ring S with Kdim S = **n** can be generated by n + 1 elements, (see 2). However, this is not always the best possible, as is shown by the following example. Define the n-th Weyl algebra, A_n, over a field k of characteristic zero, to be the associative k-algebra with 1 generated by the 2n indeterminates $x_1, x_2, \ldots, x_n, y_1, \ldots, y_n$, subject to the relations $x_i y_j - y_j x_i = \delta_{ij}$ and $x_i x_j - x_j x_i = 0 = y_i y_j - y_j y_i$. Then it is proved in 3 that A_n satisfies c) and d) above and the following weaker versions of a) and b).

a') Given $d \neq 0 \in A_n$ then there exist f and g in A_n such that $A_n = fdA_n + gdA_n$.

b') Any fin. gen. torsion A_n-module is a homomorphic image of a projective right ideal.

The questions that I would like to ask are the following.

1) Does A_n satisfy a) and b) ? (Conjecture: Yes)

2) For what classes of simple Noetherian rings do a),..,d) hold?

3) Does there exist an example of a simple Noetherian ring (or even a domain) for which none of a),..,d) hold? (Conjecture; Yes)

The proofs of the above-mentioned results about A_n utilise the fact that there exist a lot of partial quotient rings of A_n. The present known examples of simple Noetherian rings also tend to have this property. Thus , in order to prove 3) it may be necessary to solve the following question of Cozzens and Faith.

4) Find new examples of simple Noetherian rings.

REFERENCES

1) J. Cozzens and C. Faith, Simple Noetherian rings, Cambridge University Press, Cambridge, 1975.

2) J.T. Stafford, Completely faithful modules and ideals of simple Noetherian rings, Bull. London Math. Soc. 8 (1976), 168-173.

3) J.T. Stafford, Module structure of Weyl algebras, to appear.

SUBIDEALIZERS

Mark Teply

Let T be a ring with 1, and let K be a right ideal of T. A unital subring R of T that contains K as a 2-sided ideal is called a <u>subidealizer</u> of K in T. The maximal such subidealizer $S = \{t \in T | tK \subseteq K\}$ is called the <u>idealizer</u> of K in T. Assume $S \neq T$.

<u>General Question</u>: How are the properties of R and T related?

We want answers in terms of R, T, and R/K. To get good relationships we also assume that K is <u>generative</u>-- i.e. $TK = T$.

<u>Sample Result</u>: R is a (right) left max ring iff T and R/K are both (right) left max rings. [max ring = every module has a maximal submodule.]

<u>Transfer Problems</u>

1. Homological dimension.

<u>Robson (1972)</u>. If T is right Noetherian and K is semimaximal, then

rt. gl. dim S = sup{1,rt.gl.dim.T}.

<u>Goodearl (1973) and Robson (1972)</u>. If K is semimaximal, the same equality holds.

<u>Goodearl (1975)</u>. If R/K is semisimple Artinian, rt. gl. dim. T $\leq$ rt. gl. dim. R $\leq$ 1 + rt. gl. dim. T.

<u>Teply</u>. rt. gl. dim T $\leq$ rt. gl. dim. R $\leq$ 1 + rt. gl. dim R/K + rt. gl. dim. T.

Question. When do the various cases occur?

2. Gabriel Dimension.

Krause (1976). If K is semimaximal, rt. G-dim. T = rt. G-dim. S when either side exists.

Krause-Teply (to appear). If $(T/K)_T$ and $(R/K)_R$ are semi-artinian, then rt. G-dim. R = rt. G-dim. T when either side exists. If $_R(R/K)$ is semiartinian, ℓ. G-dim. R = ℓ. G-dim. T when either side exists.

Teply. If rt. G-dim. T and G-dim.$(R/K)_R$ are finite, rt. G-dim. R ≤ rt. G-dim. T + G-dim.$(R/K)_R$ - 1. If G-dim.$_R(R/K)$ and ℓ. G-dim. T are finite, then max{G-dim.$_R(R/K)$, ℓ. G-dim. T} ≤ ℓ. G-dim. R ≤ ℓ. G-dim. T + G-dim.$_R(R/K)$ - 1 provided any one of these terms exists. Inequalities are also known for the infinite ordinal cases. The extremes can occur.

Question. When do the various cases occur?

3. Krull Dimension (in the sense of Gordon and Robson).

Robson (1972). If T is right Noetherian and $(T/K)_R$ has finite length, then rt. K-dim. R = rt. K-dim. T.

Krause (1976). If K is semimaximal, then rt. K-dim. S = rt.K-dim.T when either side exists.

Krause-Teply (to appear). If $(T/K)_R$ has finite length, then rt. K-dim. R = rt. K-dim. T when either side exists.

Teply. If K-dim.$(T/K)_R$ exists and $(T/K)_T$ has finite length, then rt. K-dim. T ≤ rt. K-dim. R ≤ K-dim.$(T/K)_R$ + rt. K-dim. T provided any one term exists.

Question. When does each case occur?

4. V-rings [= Simple modules are injective.]
If T is a left (right) V-ring, when can R be a left (right) V-ring? In particular, does the V-ring property ever transfer from T to R when T is an integral domain?

Added in Dec. 1977. We have recently learned that the inequalities listed on the Gabriel dimension and the Krull dimension have also been obtained by F. Hansen.

EQUIVALENCE OF MATRICES, PRIME RINGS, NUMBER OF GENERATORS, STATE SPACES

Robert Warfield

1. If A is a matrix over a ring R, we let m(A) be the module determined by A--the cokernel of the corresponding homomorphism of free modules. If R is a (noncommutative) principal ideal domain, it is not necessarily true that if A and B are matrices of the same size with m(A) ≅ m(B) that A and B are equivalent, [Levy and Robson, Matrices and Pairs of Modules, J. Alg. 29 (1974), 427-454]. However, the only known counterexamples are matrices of rank one. Are there counterexamples of higher rank?

2. If R is a local serial ring (i.e. a local ring for which every f.g. right ideal is principal and every f.g. left ideal is principal), then any matrix over R is equivalent to a diagonal matrix. [R. Warfield, Serial rings and finitely presented modules, J. Alg. 37(1975), 187-222]. Such an R is not necessarily an elementary divisor ring in the sense of Kaplansky. [T.A.M.S., 66(1949)]. Is the diagonal form canonical? That is, are the diagonal entries unique up to associates? This is equivalent to the question of whether the Krull Schmidt theorem applies to decompositions of finitely presented R-modules.

3. If R is a semilocal (i.e. R/J(R) is Artinian) Bezout ring (i.e. finitely generated one sided ideals are principal), is every matrix over R equivalent to a diagonal matrix?

4. In 1969 [Pacific J. Math. 28, 699-719] I was able to show that a finitely presented module over a Prüfer domain was a summand of a direct sum of cyclics by using the fact that it is true locally, and using the Auslander Goldman [T.A.M.S. 97(1960), 1-24] formula $\mathrm{Hom}(A,B)_M = \mathrm{Hom}(A_M,B_M)$. One would like to proceed similarly in the noncommutative case. Here, a Prüfer prime ring is a prime semihereditary Goldie ring over which every nonzero finitely generated projective module is a generator, and one would like to prove that every finitely presented module is a summand of a direct sum of cyclics. (For Prüfer algebras over a Prüfer domain, this was done by Dale Miller in his thesis, (University of Washington, 1976). Assuming that there are enough two-sided ideals, one would like to use the localizations, which do exist. For any maximal ideal M such that R/M is Artinian, M is localizable, and R_M is a semilocal Bezout ring. Assuming, then, that one can solve the previous problem, one would like to do this one as in the commutative case. For this, one needs for noncommutative localization, some analogue of the Auslander Goldman result. As it stands, it doesn't make any sense, since Hom(A,B) is not a module. However, it should be possible to formulate the parts of it which are used in some meaningful

way in the noncommutative case. For example, if there is an isomorphism from A_M to B_M, where A is finitely presented, is there a homomorphism from A to B whose kernel and cokernel localize to zero? (This is with the assumption that M is a localizable maximal ideal.)

5. Dale Miller [Semihereditary Prime Rings, to appear] has shown that if A is an order over a commutative domain in the classical sense (i.e. a prime Goldie ring finite as a module over its center), then if A is semihereditary, the center of A is a Prüfer domain, and every finitely presented A-module is a summand of a direct sum of a free module and a direct sum of torsion cyclic modules. Conversely, [op. cit. 1.27], if A is coherent and every finitely presented module has this property then A is semihereditary. Is coherence necessary as an hypothesis for this last result?

6. An argument due to Faith shows that if R is a semiperfect ring in which every ideal which is f.g. as an ideal is projective on the right, then R is semihereditary. Semiperfect prime Goldie rings with this property on each side are precisely the prime serial rings [J. Alg. 37(1975)] for which every f.g. ideal is f.g. on each side. Do these rings have any special structure? The kind of structure which occurs in the Noetherian case [op. cit. section 5, and Michler, J. Alg. 13(1969), 327-344] seems possible. In the case where every nonzero f.g. ideal is actually a progenerator on each side,

R is a full matrix ring over a (generally noncommutative) valuation domain in which every one-sided ideal is two-sided, [unpublished].

7. For any ring A, the group $K_0(A)$ can be made into a "weakly" partially ordered group by taking the positive cone to be the elements which "come from" honest projective modules. (In general, one can have $x \geq y$ and $y \geq x$ while $x \neq y$, but this cannot happen in the presence of any reasonable finiteness condition.) A <u>state</u> on $K_0(A)$ is an order preserving homomorphism s: $K_0(A) \longrightarrow R$, (where R is the ring of real numbers) satisfying $s(A) = 1$. The set of such states is the "state space" of A, and is a compact, convex set. What compact convex sets arise in this way? (For von Neumann regular rings, see [K. R. Goodearl and D. Handelman, Rank Functions and K_0 of Regular Rings, J. Pure and Appl. Alg. 7 (1976), 195-216] and [K. R. Goodearl, Algebraic Representations of Choquet Simplexes, J. Pure and Appl. Alg. 11 (1977), 111-130].)

8. If S is a commutative ring of Krull dimension k, R a module-finite S-algebra, and A and B finitely generated modules with A finitely presented, and for each maximal ideal M of S there is an epimorphism $A_M \to B_M$, then Roger Wiegand has shown that there is an epimorphism $A^{k+2} \to B$, [private communication, August 1977]. If S is Noetherian, and in certain other cases, this can be improved to an epimorphism $A^{k+1} \to B$ [R. Warfield, unpublished]. Wiegand's result includes his result with Vasconcelos [to appear] that over a commutative ring of Krull dimension k, a finitely generated module locally

generated by n elements is generated by at most n(d+2) elements. In lots of cases, (e.g. if the commutative ring has only a finite number of minimal primes), they lower this estimate to n(d+1). Can this be done for finite algebras over these rings, and can the corresponding statement about mappings of finitely presented modules be proved in the same generality?

9. Heitman [Pacific J. 62(1976), 117-126] has shown that a commutative ring of Krull dimension k has k+2 in the stable range. Is this true for module-finite algebras over these rings?

10. In my paper in these proceedings is a statement of a theorem (to be published in detail elsewhere) concerning the number of generators of a finitely generated module over a Noetherian right fully bounded ring. This theorem is a generalization of the Forster-Swan theorem. Can a version of this theorem be proved for Noetherian rings in general (or with a milder hypothesis, such as Stafford's "ideal invariance")?

Vol. 551: Algebraic K-Theory, Evanston 1976. Proceedings. Edited by M. R. Stein. XI, 409 pages. 1976.

Vol. 552: C. G. Gibson, K. Wirthmüller, A. A. du Plessis and E. J. N. Looijenga. Topological Stability of Smooth Mappings. V, 155 pages. 1976.

Vol. 553: M. Petrich, Categories of Algebraic Systems. Vector and Projective Spaces, Semigroups, Rings and Lattices. VIII, 217 pages. 1976.

Vol. 554: J. D. H. Smith, Mal'cev Varieties. VIII, 158 pages. 1976.

Vol. 555: M. Ishida, The Genus Fields of Algebraic Number Fields. VII, 116 pages. 1976.

Vol. 556: Approximation Theory. Bonn 1976. Proceedings. Edited by R. Schaback and K. Scherer. VII, 466 pages. 1976.

Vol. 557: W. Iberkleid and T. Petrie, Smooth S^1 Manifolds. III, 163 pages. 1976.

Vol. 558: B. Weisfeiler, On Construction and Identification of Graphs. XIV, 237 pages. 1976.

Vol. 559: J.-P. Caubet, Le Mouvement Brownien Relativiste. IX, 212 pages. 1976.

Vol. 560: Combinatorial Mathematics, IV, Proceedings 1975. Edited by L. R. A. Casse and W. D. Wallis. VII, 249 pages. 1976.

Vol. 561: Function Theoretic Methods for Partial Differential Equations. Darmstadt 1976. Proceedings. Edited by V. E. Meister, N. Weck and W. L. Wendland. XVIII, 520 pages. 1976.

Vol. 562: R. W. Goodman, Nilpotent Lie Groups: Structure and Applications to Analysis. X, 210 pages. 1976.

Vol. 563: Séminaire de Théorie du Potentiel. Paris, No. 2. Proceedings 1975–1976. Edited by F. Hirsch and G. Mokobodzki. VI, 292 pages. 1976.

Vol. 564: Ordinary and Partial Differential Equations, Dundee 1976. Proceedings. Edited by W. N. Everitt and B. D. Sleeman. XVIII, 551 pages. 1976.

Vol. 565: Turbulence and Navier Stokes Equations. Proceedings 1975. Edited by R. Temam. IX, 194 pages. 1976.

Vol. 566: Empirical Distributions and Processes. Oberwolfach 1976. Proceedings. Edited by P. Gaenssler and P. Révész. VII, 146 pages. 1976.

Vol. 567: Séminaire Bourbaki vol. 1975/76. Exposés 471–488. IV, 303 pages. 1977.

Vol. 568: R. E. Gaines and J. L. Mawhin, Coincidence Degree, and Nonlinear Differential Equations. V, 262 pages. 1977.

Vol. 569: Cohomologie Etale SGA $4\frac{1}{2}$. Séminaire de Géométrie Algébrique du Bois-Marie. Edité par P. Deligne. V, 312 pages. 1977.

Vol. 570: Differential Geometrical Methods in Mathematical Physics, Bonn 1975. Proceedings. Edited by K. Bleuler and A. Reetz. VIII, 576 pages. 1977.

Vol. 571: Constructive Theory of Functions of Several Variables, Oberwolfach 1976. Proceedings. Edited by W. Schempp and K. Zeller. VI, 290 pages. 1977

Vol. 572: Sparse Matrix Techniques, Copenhagen 1976. Edited by V. A. Barker. V, 184 pages. 1977.

Vol. 573: Group Theory, Canberra 1975. Proceedings. Edited by R. A. Bryce, J. Cossey and M. F. Newman. VII, 146 pages. 1977.

Vol. 574: J. Moldestad, Computations in Higher Types. IV, 203 pages. 1977.

Vol. 575: K-Theory and Operator Algebras, Athens, Georgia 1975. Edited by B. B. Morrel and I. M. Singer. VI, 191 pages. 1977.

Vol. 576: V. S. Varadarajan, Harmonic Analysis on Real Reductive Groups. VI, 521 pages. 1977.

Vol. 577: J. P. May, E_∞ Ring Spaces and E_∞ Ring Spectra. IV, 268 pages. 1977.

Vol. 578: Séminaire Pierre Lelong (Analyse) Année 1975/76. Edité par P. Lelong. VI, 327 pages. 1977.

Vol. 579: Combinatoire et Représentation du Groupe Symétrique, Strasbourg 1976. Proceedings 1976. Edité par D. Foata. IV, 339 pages. 1977.

Vol. 580: C. Castaing and M. Valadier, Convex Analysis and Measurable Multifunctions. VIII, 278 pages. 1977.

Vol. 581: Séminaire de Probabilités XI, Université de Strasbourg. Proceedings 1975/1976. Edité par C. Dellacherie, P. A. Meyer et M. Weil. VI, 574 pages. 1977.

Vol. 582: J. M. G. Fell, Induced Representations and Banach *-Algebraic Bundles. IV, 349 pages. 1977.

Vol. 583: W. Hirsch, C. C. Pugh and M. Shub, Invariant Manifolds. IV, 149 pages. 1977.

Vol. 584: C. Brezinski, Accélération de la Convergence en Analyse Numérique. IV, 313 pages. 1977.

Vol. 585: T. A. Springer, Invariant Theory. VI, 112 pages. 1977.

Vol. 586: Séminaire d'Algèbre Paul Dubreil, Paris 1975–1976 (29ème Année). Edited by M. P. Malliavin. VI, 188 pages. 1977.

Vol. 587: Non-Commutative Harmonic Analysis. Proceedings 1976. Edited by J. Carmona and M. Vergne. IV, 240 pages. 1977.

Vol. 588: P. Molino, Théorie des G-Structures: Le Problème d'Equivalence. VI, 163 pages. 1977.

Vol. 589: Cohomologie l-adique et Fonctions L. Séminaire de Géométrie Algébrique du Bois-Marie 1965–66, SGA 5. Edité par L. Illusie. XII, 484 pages. 1977.

Vol. 590: H. Matsumoto, Analyse Harmonique dans les Systèmes de Tits Bornologiques de Type Affine. IV, 219 pages. 1977.

Vol. 591: G. A. Anderson, Surgery with Coefficients. VIII, 157 pages. 1977.

Vol. 592: D. Voigt, Induzierte Darstellungen in der Theorie der endlichen, algebraischen Gruppen. V, 413 Seiten. 1977.

Vol. 593: K. Barbey and H. König, Abstract Analytic Function Theory and Hardy Algebras. VIII, 260 pages. 1977.

Vol. 594: Singular Perturbations and Boundary Layer Theory, Lyon 1976. Edited by C. M. Brauner, B. Gay, and J. Mathieu. VIII, 539 pages. 1977.

Vol. 595: W. Hazod, Stetige Faltungshalbgruppen von Wahrscheinlichkeitsmaßen und erzeugende Distributionen. XIII, 157 Seiten. 1977.

Vol. 596: K. Deimling, Ordinary Differential Equations in Banach Spaces. VI, 137 pages. 1977.

Vol. 597: Geometry and Topology, Rio de Janeiro, July 1976. Proceedings. Edited by J. Palis and M. do Carmo. VI, 866 pages. 1977.

Vol. 598: J. Hoffmann-Jørgensen, T. M. Liggett et J. Neveu, Ecole d'Eté de Probabilités de Saint-Flour VI – 1976. Edité par P.-L. Hennequin. XII, 447 pages. 1977.

Vol. 599: Complex Analysis, Kentucky 1976. Proceedings. Edited by J. D. Buckholtz and T. J. Suffridge. X, 159 pages. 1977.

Vol. 600: W. Stoll, Value Distribution on Parabolic Spaces. VIII, 216 pages. 1977.

Vol. 601: Modular Functions of one Variable V, Bonn 1976. Proceedings. Edited by J.-P. Serre and D. B. Zagier. VI, 294 pages. 1977.

Vol. 602: J. P. Brezin, Harmonic Analysis on Compact Solvmanifolds. VIII, 179 pages. 1977.

Vol. 603: B. Moishezon, Complex Surfaces and Connected Sums of Complex Projective Planes. IV, 234 pages. 1977.

Vol. 604: Banach Spaces of Analytic Functions, Kent, Ohio 1976. Proceedings. Edited by J. Baker, C. Cleaver and Joseph Diestel: VI, 141 pages. 1977.

Vol. 605: Sario et al., Classification Theory of Riemannian Manifolds. XX, 498 pages. 1977.

Vol. 606: Mathematical Aspects of Finite Element Methods. Proceedings 1975. Edited by I. Galligani and E. Magenes. VI, 362 pages. 1977.

Vol. 607: M. Métivier, Reelle und Vektorwertige Quasimartingale und die Theorie der Stochastischen Integration. X, 310 Seiten. 1977.

Vol. 608: Bigard et al., Groupes et Anneaux Réticulés. XIV, 334 pages. 1977.

Vol. 609: General Topology and Its Relations to Modern Analysis and Algebra IV. Proceedings 1976. Edited by J. Novák. XVIII, 225 pages. 1977.

Vol. 610: G. Jensen, Higher Order Contact of Submanifolds of Homogeneous Spaces. XII, 154 pages. 1977.

Vol. 611: M. Makkai and G. E. Reyes, First Order Categorical Logic. VIII, 301 pages. 1977.

Vol. 612: E. M. Kleinberg, Infinitary Combinatorics and the Axiom of Determinateness. VIII, 150 pages. 1977.

Vol. 613: E. Behrends et al., L^p-Structure in Real Banach Spaces. X, 108 pages. 1977.

Vol. 614: H. Yanagihara, Theory of Hopf Algebras Attached to Group Schemes. VIII, 308 pages. 1977.

Vol. 615: Turbulence Seminar, Proceedings 1976/77. Edited by P. Bernard and T. Ratiu. VI, 155 pages. 1977.

Vol. 616: Abelian Group Theory, 2nd New Mexico State University Conference, 1976. Proceedings. Edited by D. Arnold, R. Hunter and E. Walker. X, 423 pages. 1977.

Vol. 617: K. J. Devlin, The Axiom of Constructibility: A Guide for the Mathematician. VIII, 96 pages. 1977.

Vol. 618: I. I. Hirschman, Jr. and D. E. Hughes, Extreme Eigen Values of Toeplitz Operators. VI, 145 pages. 1977.

Vol. 619: Set Theory and Hierarchy Theory V, Bierutowice 1976. Edited by A. Lachlan, M. Srebrny, and A. Zarach. VIII, 358 pages. 1977.

Vol. 620: H. Popp, Moduli Theory and Classification Theory of Algebraic Varieties. VIII, 189 pages. 1977.

Vol. 621: Kauffman et al., The Deficiency Index Problem. VI, 112 pages. 1977.

Vol. 622: Combinatorial Mathematics V, Melbourne 1976. Proceedings. Edited by C. Little. VIII, 213 pages. 1977.

Vol. 623: I. Erdelyi and R. Lange, Spectral Decompositions on Banach Spaces. VIII, 122 pages. 1977.

Vol. 624: Y. Guivarc'h et al., Marches Aléatoires sur les Groupes de Lie. VIII, 292 pages. 1977.

Vol. 625: J. P. Alexander et al., Odd Order Group Actions and Witt Classification of Innerproducts. IV, 202 pages. 1977.

Vol. 626: Number Theory Day, New York 1976. Proceedings. Edited by M. B. Nathanson. VI, 241 pages. 1977.

Vol. 627: Modular Functions of One Variable VI, Bonn 1976. Proceedings. Edited by J.-P. Serre and D. B. Zagier. VI, 339 pages. 1977.

Vol. 628: H. J. Baues, Obstruction Theory on the Homotopy Classification of Maps. XII, 387 pages. 1977.

Vol. 629: W. A. Coppel, Dichotomies in Stability Theory. VI, 98 pages. 1978.

Vol. 630: Numerical Analysis, Proceedings, Biennial Conference, Dundee 1977. Edited by G. A. Watson. XII, 199 pages. 1978.

Vol. 631: Numerical Treatment of Differential Equations. Proceedings 1976. Edited by R. Bulirsch, R. D. Grigorieff, and J. Schröder. X, 219 pages. 1978.

Vol. 632: J.-F. Boutot, Schéma de Picard Local. X, 165 pages. 1978.

Vol. 633: N. R. Coleff and M. E. Herrera, Les Courants Résiduels Associés à une Forme Méromorphe. X, 211 pages. 1978.

Vol. 634: H. Kurke et al., Die Approximationseigenschaft lokaler Ringe. IV, 204 Seiten. 1978.

Vol. 635: T. Y. Lam, Serre's Conjecture. XVI, 227 pages. 1978.

Vol. 636: Journées de Statistique des Processus Stochastiques, Grenoble 1977, Proceedings. Edité par Didier Dacunha-Castelle et Bernard Van Cutsem. VII, 202 pages. 1978.

Vol. 637: W. B. Jurkat, Meromorphe Differentialgleichungen. VII, 194 Seiten. 1978.

Vol. 638: P. Shanahan, The Atiyah-Singer Index Theorem, An Introduction. V, 224 pages. 1978.

Vol. 639: N. Adasch et al., Topological Vector Spaces. V, 125 pages. 1978.

Vol. 640: J. L. Dupont, Curvature and Characteristic Classes. X, 175 pages. 1978.

Vol. 641: Séminaire d'Algèbre Paul Dubreil, Proceedings Paris 1976–1977. Edité par M. P. Malliavin. IV, 367 pages. 1978.

Vol. 642: Theory and Applications of Graphs, Proceedings, Michigan 1976. Edited by Y. Alavi and D. R. Lick. XIV, 635 pages. 1978.

Vol. 643: M. Davis, Multiaxial Actions on Manifolds. VI, 141 pages. 1978.

Vol. 644: Vector Space Measures and Applications I, Proceedings 1977. Edited by R. M. Aron and S. Dineen. VIII, 451 pages. 1978.

Vol. 645: Vector Space Measures and Applications II, Proceedings 1977. Edited by R. M. Aron and S. Dineen. VIII, 218 pages. 1978.

Vol. 646: O. Tammi, Extremum Problems for Bounded Univalent Functions. VIII, 313 pages. 1978.

Vol. 647: L. J. Ratliff, Jr., Chain Conjectures in Ring Theory. VIII, 133 pages. 1978.

Vol. 648: Nonlinear Partial Differential Equations and Applications, Proceedings, Indiana 1976–1977. Edited by J. M. Chadam. VI, 206 pages. 1978.

Vol. 649: Séminaire de Probabilités XII, Proceedings, Strasbourg, 1976–1977. Edité par C. Dellacherie, P. A. Meyer et M. Weil. VIII, 805 pages. 1978.

Vol. 650: C*-Algebras and Applications to Physics. Proceedings 1977. Edited by H. Araki and R. V. Kadison. V, 192 pages. 1978.

Vol. 651: P. W. Michor, Functors and Categories of Banach Spaces. VI, 99 pages. 1978.

Vol. 652: Differential Topology, Foliations and Gelfand-Fuks-Cohomology, Proceedings 1976. Edited by P. A. Schweitzer. XIV, 252 pages. 1978.

Vol. 653: Locally Interacting Systems and Their Application in Biology. Proceedings, 1976. Edited by R. L. Dobrushin, V. I. Kryukov and A. L. Toom. XI, 202 pages. 1978.

Vol. 654: J. P. Buhler, Icosahedral Golois Representations. III, 143 pages. 1978.

Vol. 655: R. Baeza, Quadratic Forms Over Semilocal Rings. VI, 199 pages. 1978.

Vol. 656: Probability Theory on Vector Spaces. Proceedings, 1977. Edited by A. Weron. VIII, 274 pages. 1978.

Vol. 657: Geometric Applications of Homotopy Theory I, Proceedings 1977. Edited by M. G. Barratt and M. E. Mahowald. VIII, 459 pages. 1978.

Vol. 658: Geometric Applications of Homotopy Theory II, Proceedings 1977. Edited by M. G. Barratt and M. E. Mahowald. VIII, 487 pages. 1978.

Vol. 659: Bruckner, Differentiation of Real Functions. X, 247 pages. 1978.

Vol. 660: Equations aux Dérivée Partielles. Proceedings, 1977. Edité par Pham The Lai. VI, 216 pages. 1978.

Vol. 661: P. T. Johnstone, R. Paré, R. D. Rosebrugh, D. Schumacher, R. J. Wood, and G. C. Wraith, Indexed Categories and Their Applications. VII, 260 pages. 1978.

Vol. 662: Akin, The Metric Theory of Banach Manifolds. XIX, 306 pages. 1978.

Vol. 663: J. F. Berglund, H. D. Junghenn, P. Milnes, Compact Right Topological Semigroups and Generalizations of Almost Periodicity. X, 243 pages. 1978.

Vol. 664: Algebraic and Geometric Topology, Proceedings, 1977. Edited by K. C. Millett. XI, 240 pages. 1978.

Vol. 665: Journées d'Analyse Non Linéaire. Proceedings, 1977. Edité par P. Bénilan et J. Robert. VIII, 256 pages. 1978.

Vol. 666: B. Beauzamy, Espaces d'Interpolation Réels: Topologie et Géometrie. X, 104 pages. 1978.

Vol. 667: J. Gilewicz, Approximants de Padé. XIV, 511 pages. 1978.

Vol. 668: The Structure of Attractors in Dynamical Systems. Proceedings, 1977. Edited by J. C. Martin, N. G. Markley and W. Perrizo. VI, 264 pages. 1978.

Vol. 669: Higher Set Theory. Proceedings, 1977. Edited by G. H. Müller and D. S. Scott. XII, 476 pages. 1978.